THE ART OF *Energy*

THE ART OF *Energy*

YOUR BLUEPRINT FOR GREAT ENERGY

GURUTEJ KHALSA

The Art of Energy

by Gurutej Khalsa
1310 Carmona Ave.
Los Angeles CA 90016
USA

ISBN 9-798691-432606

CONTENTS

About the Author xii

Introduction xiii

Check in, Checkup 1

Fake Energy 4

Average or Successful – You Decide 7

An open mind and the willingness to shift 11

There are only a few major components to this program... 12

The Truth about Sleep 12

The Truth about Waking Up Properly 13

The Truth about the Foods You Eat 13

The Truth about Meditation 13

The Truth about Exercise 14

The Truth about Breath—Your Breath 14

Testimonials 15

Testimonial 20

Sleep 21

Wake up 21

Food 22

Exercise 22

Meditate 22

The 7 Steps to Success 23

Testimonial 24

Testimonial 30

The Cycle of Our Habits 31

The Truth about Sleep 39

Claw Dance 40

Single Nostril Breathing..40
Sitali Breath, The Night or Relaxing Breath......................41
Meditation 1. Before going to bed...................................46
Meditation 2. Creating calm and rejuvenating the glands (more advanced)..47
Things that can assist your sleep......................................47
Where you sleep and what you sleep on matters...............50
There is a secret to 10-minute naps and the best hours for sleeping and naps...52
Four-Part Forward Bend..53
Spinal Rolls...53
Best Hours to Sleep and More...54
Meditation to Relax You..56
What You Do During the Day Matters...........................57
Desk Bowing...58
Your Sleep Problems Solved..59

The Truth About Waking Up Properly........................... 62
We have 10 bodies..62
Now you are ready for the bathroom...............................66
Taking a cold shower: oiling your body first wakes you up and clears your head.................................67
Next is meditation..68
Meditation...69
What you eat matters...69

The Truth about the Foods You Eat............................ 72
You need to eat real foods, not things masquerading as food. Think green..72
LBN-Commentary by Jackie Keller.................................73

Sodas' Disastrous Side Effects 75
This article is even scarier or let's say hopefully more awakening..77
A New Zealand woman dies from drinking more than two gallons of soda a day..77
Energy Drinks 77
Sugar 79
Eat light, easily digestible foods..80
Article on Food and Obesity from Huffington Post..........80
Eat as much organic food as possible..82
How to sprout any seed or bean..82
Raw Hummus Recipe..83
Green Dip 84
Vegan Dark Chocolate Mousse..84
What you drink during the day makes a huge difference in your true sleep 87
Eating small meals often works best for most 88
To eat or not to eat before you sleep 89

The Truth about Meditation 90
We all have a monkey mind..90
Here is a study published in Psychology Today.................91
Meditation calms the mind..93
Meditate on This: The Practice Can Heal You in Less Than 11 Hours. —Stephan A. Schwartz..........................93
Sipping Breath Meditation..94
Try this next meditation:
When you don't know what to do 94
Here is another instant calm 95

Meditation for peace and calm..96

The Truth about Exercise 98

Renew Exercise 98

Our Sedentary Society 98

Hand Dusting...103

Lasso Pose...103

Claw Dance 104

The Truth about Breath—Your Breath **105**

Three Ways Your Breathing Can Improve 105

Our breathing is an automatic response 107

There is not just one breath for all occasions 109

Dragon Breath 110

Having an army of breaths for all situations is necessary..111

Stress robs you of your breath 111

Breathing is the fastest road to energy, always. It's the original Red Bull...111

Breath of Fire...112

Here is another de-stressor during the day.......................112

Four-Part Breath: Miracle Breath.....................................113

Sitali Breath 114

O Breath...114

Taking Stock of Your New Life **116**

Final Tip: Summation **118**

Note to the Reader

This book contains useful information. However, the author and publisher do not hold any liability.

"To everyone who's ever wished they had more time on their hands to do what they've always wanted to do in life and the energy to do it."

-Gurutej Khalsa

About the Author

For over 50 years, Gurutej has been living her destiny and continues to awaken the spirit of all she touches. Gurutej emerged as a founding practitioner of Kundalini. Gurutej is now considered a foremost authority on Kundalini Yoga and internationally recognized as one of a handful of Kundalini Yoga Masters and more.

Through yoga, chanting, meditation, healing, and proven stress and health management systems, she teaches people how to connect to their higher consciousness. As a prolific writer, Gurutej is an author to 3 popular books including: A Slice of the Beloved: Yoga for Relationships, The Moon She Rocks You: Revealing the Secrets of Women's Inner Emotions, The 13th Month: How to Get an Extra 29 Days Each Year, and The Art Of Energy, Your Blueprint for Great Energy. She has also developed the transformative powerful online Academy Empowered Energy Academy. To support and uplift people through her 11 online courses.

Gurutej has the depth, breadth, insight, and training to help you with all those issues on love, success, relationships, energy, health, balance, and much more. You can work with Gurutej privately.

When not traveling the world and teaching, Gurutej can be found at The Blessings Center in Los Angeles. You can also enjoy her blog or suite of online books, programs, classes and products at gurutej.com

INTRODUCTION

What if a magic genie came down and gave you two extra hours each and every day? Would you accept the gift? Would you be ecstatic? How would you like to sleep less yet FEEL BETTER?

This book and program will give you tools to do just that. It is for those who want more and are fearless about what they are willing to do to achieve great true success. Are you more curious or more resistant? If you are resistant and this is not what you want, please pass this book on to someone you know who really wants a more powerful, dynamic and fulfilling life and is willing to do what it takes to have that life.

Ask yourself: What would make you be willing to sleep less? I want you to look at these numbers and see if they inspire you. *If you gained two waking hours per day—that is 14 hours a week, 60 hours a month, and a glorious 728 hours a year—then you would gain over a month each year.* What a free gift that could be, yes? If you take me up on this process, it can be yours, as well as much more health and wealth.

Just think what you could do with that time—one extra month in your life each and every year you do this.

Think…

- how much more money you would make.
- how much more time you would have with the people you love.
- how much more time you would have to pursue your passions doing what you love.

But before I even get into this, I want to tell you my story and what inspired this book.

I was taking an intensive course for authors. The person teaching the course, Steve Harrison, asked us to fill out a very extensive form which he looked at before our one interview with him. Of all I had done in my life (and it is a very wide-ranging list), he was most impressed by the fact that I had slept for only 3-4 hours each night for 35 years. He asked if I would like to write a book about how I did this, and the benefits it would give others if they followed this system, and he wanted more hours in his day. He was also impressed by the fact that I was 63 and still looked amazing (as others say) and feel great (see my pictures at www.gurutej.com). I can do handstands, headstands, rollerblade hills, and much more. I also have a very clear, concise, and productive mind. I imagine if I had looked haggard and spent, he would not have been as impressed with the 35 years I spent getting only 3-4 hours of sleep each night. Don't let this scare you. I'm not asking you to do that. I was part of a consciousness experiment in the 70s. I will tell you about that soon.

Sleep is not the same for all cultures. What we don't realize is that other cultures have very different dynamics with sleep than we do. We know that some cultures, notably the Italians and the Spanish, have siestas (adult nap time). Some of you may think it's silly, but most adults I know would say, "How delightful." These cultures tend to have dinner at 10 pm and go to sleep around midnight or later. The Japanese stay out until the wee hours and, yes, drink like fish, then wake up early doing their Tai Chi or some other form of exercise before the sun rises. People in India also have the "late to bed, early to rise" policy, but based on something a little different. They believe, as many mystics and sages do, that getting up to pray or meditate as early as 3 or 4 am is the best time to commune with the self and the higher self.

Christian mystics also follow the same timeframe for the same reasons. There are fewer distractions in these early morning hours for most of us, making it easier to connect with our deeper self, and yes, with the vastness of the universe. This is the time that you get great downloads. As you clear your mind and your subconscious, you make room for the broader concepts to come to your mind that can aid you in your personal and business life.

Back to my story: I was a major night owl at university doing lots of extracurricular things, including lots and lots of drugs. I loved the state of being high and feeling very connected, but I didn't like the coming down or the emotions that followed. Then I met this guy who impressed me as a very happy person. I got to find out his secrets for being so happy. He loved to cook. This was in 1969 and he was a vegetarian. I never knew there was such a thing. I was from Missouri, the "Show Me" state—show me the cow, show me the pig, show me the chicken. We ate meat three times a day at home. But I must admit I only liked mine charred and disguised. So, when I heard the vegetarian thing and saw how he loved to cook and how great his food tasted and how I felt after eating it, I jumped on the bandwagon, never to look back. I know this is not true for everyone, but this is my story. You can sit with your meat idolatry if you like. (Eating consciously raised meat a couple of times a week is not a bad thing.) He figured, "Wow, what an easy convert," so he showed me that he was doing yoga in his closet. It gives a whole new meaning to "coming out of the closet," doesn't it? I liked what I saw but I wanted to experience yoga for real, and not just from a book. Remember, this was 1969 and almost no one had even heard of yoga.

Then a great thing happened. My university speech class teacher gave us the assignment of doing a speech on something we were interested in but knew very little about, though he actually said, *know nothing* about. A light turned on in my head—why not yoga? So, I went to the library, something I didn't do a lot of in those days. There, I found manuscripts (I wonder why these are not called *womanuscripts*?) on yoga. These were so rare that I was not allowed to take them out of the library. These books showed great little skinny men (but no women) in what looked like diapers, and they had truly bright shiny eyes. They explained the whys, the great effects of yoga, and showed different poses. I liked it, so I did the same when I delivered my speech, showing some of the poses I could do. I could actually do many of the poses, but that didn't strike me as odd in those days. Like any good public speech, I had 14 students come up to the front right after my speech, asking where they could do this thing called yoga. I didn't have an answer, so I said, "Let's let each other know if we find out." There was no text nor cellphones in the 70s.

In May of 1970, someone ran into me where we were and said, "There's yoga in the park." These words changed my life forever. I went. I fell in love and have been doing it ever since. It was recommended that we get up early and make it to the EARLY morning yoga and meditation at 3:45 am. Yes A.M.! Since I went to bed at 2 am, it sounded insane to me, but I wanted to see what it was about. Let's say my curiosity got the best of me. I showed up and liked it so much that it turned my whole life around. My teacher's mantra was, "Eat less, sleep less, and meditate more." Not exactly the American dream but still, the dream was to build things. In our case, we were building communities, businesses, and yoga centers all over the world. There was neither time nor support for sleeping a lot. So, for 35 years, I slept 3 to 4 hours a night, meditated, did yoga, and kept a vegetarian or vegan diet. Oh, did I say I was running most of the communities in Canada at age 22? Previously, the biggest community I had been part of was a family of eight, a boarding school, a university—and I had not studied how any of these ran. It was an on-the-job training and the entrepreneurial spirit was alive and well in us all in the 60s and 70s. And today, it is back.

Why am I telling you this? *Because you too have asked how you can do more and have more energy and clarity to do what you want and need to do in your life. That is how you found this book or program.*

This program is based on what I learned. This is a distillation of all I've learned, with the imperfections taken out. You get all the goodies with the stuff that didn't work removed. Now, in order for you to venture forth, I need you to answer some qualifying questions.

—Gurutej Khalsa

Check In, Checkup

Please be truthful. This is just for you to see before you do the program. Then we can get to work on this together. Right now, it doesn't matter if some of these times are zero or 10. We will take care of that.

Please answer these questions.

How many hours a day do you sleep?

How many hours do you meditate?

How many hours do you exercise?

How many hours do you work?

How many hours do you spend with your family or friends?

How many hours do you spend doing the things you truly love—in other words, your passions?

How many hours do you waste avoiding things you don't want to do?

How many hours do you spend surfing some site on the Internet, gaming, or sitting in your favorite black hole?

Total of Hours:

How many hours did it add up to? It is likely that the number is more than the actual hours you are awake. Would you like to have those extra hours?

This is an exceptional program for exceptional people. If this is what you want, then proceed. Otherwise, turn tail and run away. Only when your passion is greater than your fear will you be able to make changes in your life. To proceed, you need to be unhappy with what is, and hungry for something better and more.

The first requirement is that you have to be hungry for those extra hours. You have to know the things you really want to do with these hours, to drive you to take the necessary actions and do what it takes. It means giving up things to get what you want.

You have to have a greater love for what you want than what you think you will lose. If you are in this group, then come with me. This is for you love.

Remember that this is what you could get for your efforts—14 extra hours each week, 60 extra hours each month, and the amazing 728 extra hours (1 extra month) each year.

Ask yourself the following questions. These too require YOUR truthful answers. Yes, there is no room for lying here or the great cold shower monster will come to claim you. Remember, no one sees your answers but you and the little devil—I mean angel—on your shoulder. This will help you come to grips with how serious you really are.

How much more money do you think you could make with 14 or more extra hours each and every week?

How much more could you create?

What fun things could you do with your family or friends?

What creative or fun things could you do that you have always wanted to do for yourself?

How could you take care of yourself better if you had more time?

How many more books could you read or courses you want to take?

How much better could you eat and take care of your health.

Fake Energy

I'm going to give you just one priceless tip that could make a difference in your energy levels for the day. *Drop the coffee—no, not the cup in your hand right now, but coffee.* Yes, that means the one cup in the morning and perhaps the others you have the rest of the day, or those little energy shots of heart-attack amounts of caffeine. What for? Because it is a bad lover—it smells great, tastes great, can even look great, but the energy it gives you is fake energy. If you don't have another cup within a very specific timeframe, that fake energy *levels you* after a few hours. It drops you hard and fast and deadens your adrenals. It takes over the work of the adrenals, keeping you awake and alert and also very much on edge. This goes for the drugs Ativan and Ritalin as well. Even though they are in another category of hyper-focus drugs, they treat you badly, very badly. Adrenals stop working because you are telling them that you have something else that will do their work for you, "thank you very much," so they stop working. You may say, "So what? They can't be that important." But guess what? They sit on top of your kidneys, and kidneys are the organs that tell you what age you really are. When the adrenals are down, your kidneys also become weak and drained. When the kidneys are weak and puny, it's draining to your pericardium, which is the sack your heart rests in. The Chinese would say that your Kidney Chi becomes weaker and weaker as you age, which is true. As you age, your muscles also become constricted. While coffee can make you feel ready for physical activity on a daily basis, it makes you uptight, very inflexible, and stiff. It is a vasoconstrictor.

Want a story? I understand I am attacking the next most sacred thing to apple pie—actually maybe more sacred than apple pie—coffee. So here is a coffee anecdote and if you have your hands over your ears and over your eyes, then come back to this when you might be ready. I have had many people who want to hear nothing about giving up the sacred cow that is coffee. Well, this lady in my class really did not want to hear anything about giving up coffee. She was not going to give it up but she was really tired of having her muscles feel so weak when she was doing yoga. She almost gave up the yoga for the coffee but decided she would give coffee up but only for 40 days, then she would reclaim it. Well, she so loved her newfound strength so much that she didn't want to lose it. She could hold her arms up indefinitely and could even do pushups. She didn't want to cry when we did arm exercises any more. "Damn," she said, "I want my coffee but I want my strength more, so I guess that means no coffee for now."

Substitute for coffee? Not Chamomile tea (that is so not my favorite), but something that has body like coffee and a little caffeine too, like Chai tea, without the sugar factor.

You can buy original Yogi Tea bags or Green Chai to make it easy, or a loose pack to make a big batch. Add a green tea of your choice and whatever milk substitute you want or drink it black. If you love it sweet, try raw organic honey, Birch Xylitol, or Stevia. But not those cute little poison packs of artificial sweeteners that stay ziplocked in your cells forever and ever, creating a reign of terror.

The solutions are simple but our habits are not so simple. They make us captive. But if you are willing to be a free agent again, then now is your time to plan your escape into a realm of true higher energy and MORE time.

So, you already know that someone who is highly successful was surprised to hear about my many years of 3 to 4 hours of sleep and gave me the shove to write this book. He said he knew many people who were looking for more hours in their day and asked if I would be willing to share my secrets. That is when I realized the TOTAL POWER of what I had been living.

Average or Successful—You Decide

Here is the average person's checklist. Mark the ones that pertain to you:

1. Do you wake up to coffee and eat badly (something you can grab quickly) or don't eat breakfast (way better)?

2. Do you feel that you are gathering parts as you go, trying to make a whole person as you run out your door in the morning?

3. Do you abstain from drinking enough water because it means you have to go to the bathroom often?

4. Is coffee or soda your main source of fluids during the day?

5. Do you snack on junk that passes as food, or do you contribute to bringing home junk food, keeping it in your desk drawers, car, or purse?

6. Do you eat meals at your desk?

7. Do you resist taking the time to exercise or do you hate it when you do?

8. Do you work long hours or bring work home, or both?

9. At night, do you treat yourself to more junk food, drugs, caffeine, alcohol or a combination of these as a reward for a grueling day?

10. Do you binge-watch TV or something on your computer to unwind because it won't talk back to you, ask you questions, engage you in conversation, and require you to interact or pay attention?

Being exhausted by the day's end, used up, spent, or done in does not make for juicy relationships.

Here is the successful person's checklist: Please mark the ones that pertain to you.

1. Gets up before sunrise to start the day by connecting with herself/himself quietly.

2. Showers, often with cold water to truly wake up and have a light breakfast.

3. Gets to work before others to have uninterrupted time at work.

4. Goes out for lunch, and, if eating with stressful people, drink (non-alcoholic drinks) and doesn't eat. She/he eats later.

5. Breathes through every task and situation during the day because it adds energy and awareness to the situation.

6. Knows what to say "yes" to and what to say "no" to during the day. Prioritizes the day.

7. Sets a deadline to leave the office and sticks to it. (mostly)
8. Is willing to change when things call for a change.
9. Let's go of the day before bed so he/she doesn't take the day to bed with.

How many do you have in the Successful Person Category, and how many in the Average Person Category?

Please go back and check them off, then add them up.

Average Person: ______________________________

Successful Person: ___________________________

Which category do you want to be in? It is a choice.

Just three things.

How would you like to sleep less and feel better? Then try these three simple things for starters. Really, there are only 3 **things you need to do to get started with this success path.**

1. Meditate for 11 minutes first thing each morning and last thing at night. This will clear your mind for the day so you can start clean and clear at night so you sleep better.

2. Switch from coffee to green tea—just try it for 40 days.

3. Eat at least 40% live foods or whole living foods. Try organic unpackaged foods, except the salad greens.

Okay, I know I said 3, but this is a bonus and takes no time or effort: *Pay attention to how you go to bed.*

What if I told you that you could sleep less, feel better, have more time and be less stressed? If this was a pill, you all would be running out to buy it. The idea that we need 8 hours of sleep is a myth. Let me tell you a story. When I go to India, I am always amazed because people there would pull their shawls around them and just go to sleep on temple floors for 3 hours and get up, ready to go. This always amazes me as I am usually very jetlagged after traveling by air for about 24 to 26 hours from LA to India. Even in the wealthiest houses, people in India only sleep for a few hours. They don't have the concept in their minds that they need 8 hours of sleep, so they are fine with much, much less. You can say, "Oh, they have less stress," but you would not say this if you have been to India. Just crossing a street is one of the most harrowing things you will ever face. Try getting between rickshaws, bicycles, cows, people, buses, cars, and trucks, not to mention mopeds with multiple people on each one. They all move at a different speed, making it an obstacle course of the most intense kind. Oh, and NO ONE stops for you. It is loud, colorful, wild, polluted, and encompassing. But the people there want to feed you. They love having you in their homes. They will share whatever they have with you even if they have almost nothing. I am not telling you that Indian food is good for you, nor the heavily sugared and caffeinated Chai. I am telling you that people in India don't have a belief system that says you need these long uninterrupted stretches of sleep.

Want another story? When I was telling a friend about this book, he relayed this experience. He said he used to meditate each morning for at least an hour and was only sleeping 4 hours at that time as well. He said he and several other people were in Lake Tahoe with Buckminster Fuller, the great mystic/architect, and they got to take long hikes together every morning because Mr. Fuller only slept 3 to 4 hours as well. All the other people slept 7 to 8 hours and missed this amazing private time with "Bucky" as he was lovingly called. What an amazing chance to spend time with such a great mind and being, all because my friend wasn't sleeping in. He got the goodies. Are you ready for your goodies?

Successful people are always willing to challenge themselves and their concepts.

Are you willing to do that?

You only need to bring 2 things to this party to get the most out of it and succeed in life.

An Open Mind and the Willingness to Shift

These things sound so easy, yet I know habits are the hardest thing to change. Are you fed up with not having enough hours in your day to do the things you love and want to be doing? Only then will you be willing to jump on this success train. All successful people utilize these two aspects. They meet the world curious and willing to see if it will benefit them. Please arm yourself with a willingness to change. Change can be scary because it is unknown, yet wonderful because of the opportunities in the unknown. Everything in life has its dos and don'ts. If you are reading this book, it means two things: you are smart and you are driven to do something that is more important to you than sleep. It's not important what that passion and drive is. I'm just acknowledging you have "the fire."

Some people cherish sleep while others want to find out how to have more waking hours. I am speaking to the latter group.

There Are Only A Few Major Components to This Program

Here are the basic tenets of what I'm going to tell you, laid out really plain and simple.

Sleep is necessary. The amounts we all need vary. You can change the amount of sleep you need with several things.

Meditation, breath, diet (the foods you eat and the ones you no longer will), and exercise are important.

Energy, your energy, and having the tools to regain that energy when you lose it throughout the day. The people in your life either give energy or take it. Knowing who is who and using the tools to release and rejuvenate yourself will prove priceless. What matters are the people you choose to be around and how you choose to deal with the ones you don't choose but still "have" to be round.

These are the things that affect your energy. The more energy you have, the less sleep you will need. How do you get that energy going? We are going to work your day backwards. Keep that mind wide open.

The Truth about Sleep

- How you go to sleep matters.
- Getting ready for sleep is an art and science.
- Where you sleep and what you sleep on matters.
- There is a secret to 10-minute naps and the best hours for sleeping and napping.
- If you can't shut your mind off, you won't sleep well.
- What you do during the day matters.
- What you eat and drink effects your sleep.

The Truth about Waking Up Properly

- Stretching in bed for 5 minutes before you get up will reset your brain.
- Oiling your body and taking a cold shower first thing wakes you up and clears your head.
- Meditating before you start your day sets you.
- Breakfast is the one most important meal of the day for some people. Find out if you are one of these people or not.
- Drinking lemon water before you eat, followed by tea, not coffee (it's Hiroshima in a cup) is a great idea for digestion.
- Body set.

The Truth about the Food You Eat

- You need to eat real foods, not things masquerading as food.
- Think green.
- Eat light, easily digestible foods.
- What you drink makes a huge difference in your sleep.
- Eating small meals often works best for most.
- Intermittent fasting is a great system set, not just for losing weight.
- To eat or not to eat before you sleep—there is no absolute law. Knowing what you need matters.
- Find out the secrets to not gain weight and feel fed.

The Truth about Meditation

- Meditation calms the mind and gives you space for answers to come to you.

- Meditation turns the volume down so clear thinking can ensue.
- All highly successful people know how to get still quickly.
- Meditation will make you more present with each situation in your life.
- Compassion.
- Meditation will give you new insights.
- Stronger Nervous system

The Truth about Exercise

- You need it.
- Choose something you like and get someone you like who will do it with you.
- Find things you can do during the day at your desk for 1 to 2 minutes. You will get some of these.
- Use tricks to get enough of it—putting a yoga mat by your desk or sofa, running the stairs, riding your bike to work, or walking the kids to school.
- Stretching is the alter ego of hard workouts. Yoga helps you last longer at any sport you may love.

The Truth about Breath—Your Breath

- Thank God it's an automatic response or we would all be dead.
- There is not just one breath for all occasions.
- Having a tool kit of breaths for all situations is necessary.
- Stress robs you of your breath.
- Breathing is always the fastest road to energy. It's the original Red Bull or Full Throttle.

Now we are ready to fill in the huge gaps one by one.

You are either curious at this point or feeling like this is all too much. You may think there is no way you can deal with all these at once. The good news is you don't have to. I want you to know about them because knowledge is power and we will set up a way for you to take them on one at a time. Pick the one that could be easiest or hardest for you. This depends on if you want to take on what is challenging you most or the back door of what is easiest. Both choices are good. Know what works best for you. ONLY choose one area to focus on and make it yours, or you may be the kind of person that says, "Give it to me hard and heavy. I want it all now because I want to succeed at all costs. If this will give me what I need to make breakthroughs and you can show me other people who have done this, then bring it on."

TESTIMONIALS:

Tom and Emily Kaplan, owners of 5 Hugo's Restaurants

How has meditation played a part in your business life?

Tom: "Meditation reflects in my business day. For me, meditation is more about viewing myself, seeing how things went the day before and understanding from a very neutral point how I could be more calm, kind, and effective in situations. It reveals where I wasn't and what I can do to rectify this today. It gives me hope that this is a new day and things that didn't work previously can work today because I have more insight."

Emily: "For me, my meditations are like revelations. [Tom's] meditations are in the form of activity while mine come in quietly. It's like I download things I have no idea I am even thinking of. I get amazing ideas on how to approach projects. I see new ways to do things in the business that turn out to be very effective and often innovative. I see more effective ways we can do things and create systems we already have in place. In meditation, I have no intention of focusing on these things—they just come through."

Tom: "They are really great innovations for the business. We would never have them unless she was meditating."

What would you say about exercise in your lives and how that affects your business life?

Tom: "I look forward to my three yoga classes a week, my bike rides, and hikes. These are my revelation times. Things come to me while I am moving my body. For me, silent exercises with or without others is the important factor. Yoga is a quiet practice, much like riding or hiking—alone, or with my wife or just a few others who can value nature and silence. This feeds both me and my business."

Emily: "I love to dance and do yoga. These open me up and make me feel happy to have a body I want to live in. It also makes me more aware of others because I am not focused on how I feel; probably because I feel great. I do what I need to do to feel the joy in life so I can deal with the problems in business from a joyous, full state."

How do you feel food impacts your day?

Emily: "I know that if I have a protein shake in the morning and nothing else, that works for me. I have lots of energy and mental clarity. A salad for lunch keeps me light so I don't bog down in the afternoon. Then I can eat a big dinner if I feel like it. I am becoming much more aware of how food *affects* me. This is making me more conscious of the choices I make. I like feeling energetic and clear. When I make funky food decisions, I lose those feelings. Not what I want."

Tom: "I want to share a story that just happened in our business. We recently opened our third restaurant, Hugo's Augora Hills, and I really pushed my partners to join the Chamber of Commerce there. What followed afterwards is interesting. The president of that chamber started frequenting our restaurant and hitting on our waitresses. They complained to me and yesterday, after my meditation, I wrote a letter of resignation to the Chamber of Commerce. I stated why Hugo's was withdrawing, and if they got a new president, we would consider rejoining. I was able to be very neutral in writing this because I did it immediately after my meditation. I brought that neutral energy into the letter. Sometimes, as the day wears on, my patience can get thin and I need to remember to take a break and spend a few minutes reconnecting because it makes such a difference in my approach."

What are the long-term effects on your business?

"We are more open to change, new ideas, not afraid to try new foods, or sources, or recipes. We are trying new events regularly. Recently, we had a Vegan Bloggers night where we feed vegan bloggers from our vegan items. They loved it and we got wonderful reviews. We are a full spectrum restaurant, meaning Omnivore to Raw; Gluten-free to Vegetarian. So, we can try many unusual and interesting events if we allow the ideas to come in. This is true expansion. I attribute this to our practice. We get expanded and it allows us to expand. We started with one restaurant and we now have five!"

Let's make this really simple. The truth is, simply changing habits is slightly more complicated than just saying, "Presto! I want to change that."

I had a client for years who was a therapist and knew how challenging it was to change habits. She saw all her clients wanting to just talk about their habits but not change them. She was truthful enough to see that she herself wanted to keep things the same. She had this priceless saying about this: "Be it ever so shitty, there is no place like home." We are at home with our habits but if you want to be more successful on all levels, more vast of brain and kinder of heart, then some sacrifices will need to be made on the altar of your better, more successful self. If you are ready for a new bigger, vaster home, try this.

You need to answer some questions first, so you know the most important place for YOU to start.

Is your brain a computer that won't turn off at night?

Do you feel tired after eating?

Do you feel tired when you get up in the morning?

Do you blast yourself awake with coffee?

Are you open to trying something new even if it seems weird?

Are you grumpy, bitchy, or short-tempered often?

Do you feel rested after sleeping?

Do you have a bedtime routine?

Do you keep food in your desk drawer, purse, briefcase, car, etc.? If yes, what kinds of food?

Do you currently meditate? If yes, how often?

a. Every day?
b. Sometimes?
c. For how long?

What kind of food or food impostors are in your daily diet? What times of day are you most prone to eating junk?

a. food in the morning
b. food from noon to 6pm
d. food at night

Do you consciously shop for your food with a food plan or just eat what is around?

Do you drink enough fluids (sodas and caffeine don't count) during the day? At night?

How often do you exercise? Every day? Once a week? Never?

How is your current energy level? High? Medium? Low?

Do you have sustained or erratic energy levels—that is, some days are good and some days are not?

From this information, I want you to see the most critical area of your life that needs to be addressed first.

How can you tell this?

Don't lie to yourself. Look at your answers and they will tell you very clearly what the most neglected part of your whole self is.

What is your most neglected part? Yes, write it down.

TESTIMONIAL:

Dr. Wondwossen Goshu

My story is totally backdoor. I met this woman, Gurutej, at a conference and we connected. She gave me a copy of her first book on relationships called "A Slice of the Beloved." It was one of the most beautiful books I ever saw or read. I was so deeply touched that I knew I needed to be close to her when I came to LA. Luckily, I was able to stay with her when I came to California to go to UCLA for a masters (producing and film making) class. On my second week, I attempted to meditate and do yoga with her. I found I was able to quiet my mind. I won't tell you meditation was easy for me because it wasn't. But I saw the kind of persons she and her partner were, so I wanted that. It made me much more aware, so I knew where and with whom to focus my time and energy with while I was here. It gave me back my power of discernment and I made better decisions about everything. I was not running around trying to get it all and do it all like many of my classmates. I just knew who I should focus on, connecting with what I wanted and needed to do.

The best part of meditation is that it is free, and you can do it whenever and wherever you are. Bringing that state of mind into my day is amazing. I just completed my doctorate in medical science. So, I am in the medical field and my love is making movies. I intend to be very successful in both areas of my life.

I need lots of energy to do this. Running and meditating are helping me achieve success in both worlds.

I want you to focus on your area of greatest neglect. Choose to focus on that first. I am going to give you three easy things you can do to bring each area back into harmony. Each of these things will be described in greater detail later in the book. This is not a popularity contest—it is about giving you more freedom in your life to do what you love and getting the results that you want and deserve in all areas of your life.

You get to focus on the first and then the others will have their turn. You are here because you want success more than comfort. You are here for one reason and one reason only—to succeed in all areas of your life. You want a better, fuller life, where you are not stealing from one part to feed the other part.

The following are just outlines of what you will get to do. Pick your weakest area and look at what is required for that area—only three things for each one, but don't try to tackle them all because that is a recipe for failure. PICK ONE AREA ONLY.

Sleep

Exercise before you get ready for bed.
Take a shower or bath.
Meditate.

Wake Up

Move your clock/phone away from the bed so you've GOT to get out of bed.

Go back to bed, breathe, and stretch in bed. Or do a couple of stretches by your bed.

Oil yourself and take a cold shower.

Drink green or white tea that you really enjoy, or try one if you have not done so yet.

FOOD

Have healthy choices at home and at work.

Eat when hungry, not when "starving," and not all the time.

Drink first—you may really be thirsty.

EXERCISE

Use my flip charts

(https://gurutejs-school.thinkific.com/courses/yoga-flip-charts) to do things at your desk to release and rejuvenate you all day long.

Commit to exercising for 20, 30, or 60 minutes daily.

Do something with your kids, partner, spouse, etc. It will be fun and will connect you more deeply with them.

MEDITATE

Do this in the morning and evening for 11 minutes to start with—11 is the Guru Number. It is the number where you can break through barriers and become wise. Guru literally means "one who takes you out of darkness into light." I say, the one who shows you all darkness and light.

Pick a meditation you are drawn to and stick with it for 40 days—the same meditation with no breaks, for 40 consecutive days.

Be grateful for each breath and use your breathing and awareness to expand you during the day.

The 7 Steps to Success

Step #1. **This is the accountability factor:** Commit to yourself and to someone else who will hold you accountable with grace and humor. You can also do this program with someone else. This increases the power because you will have a built-in accountability factor. This is the foundation for succeeding at anything.

Commitment will pay dividends. Just do it like AA—one day at a time and don't think of forever. Just string the days together and make a wonderful chain that will not bind you but free you.

Step #2. **Look at the area you are weakest in OR the easiest for you and focus on that first.** Work back from there.

Step #3. **You may feel worse before you feel better.** Detoxifying or deprivation on any level always makes us feel worse first. It gets better after the first week or two.

Remember, you get 29 extra days a year or 1 entire extra month if you do this. And you will feel better and be in amazing shape physically, mentally, and yes, spiritually.

Step #4. **Involve those around you for the best results.**

Step #5. **Do at least three new things from one area of the above list every morning and every evening. Just three will make a difference.**

Step #6. **Be willing to experiment with sleeping 1/2 hour less and supplementing meditation for that 1/2 hour.** This will give you more energy than that 1/2 hour of sleep. Then start subtracting 10 minutes a day from your sleep time—little increments each day until you to get to where you want to be so that you can do all you want and need for your business, for yourself, and for your relationships. This is Real Success.

Step #7. **Pay attention. You are smart.** Observe and pay attention to the daily, weekly, and monthly results. Remember to reward yourself for the good job you are doing. If you fall flat, pick yourself back up and get on the horse. Don't flog yourself to death and then lie back down and die. No one wins with that.

TESTIMONIAL:

Robin Sheldon, Artistry Agency LA, NY, London

I have been a business owner in a highly competitive field for over twenty years. I own a talent agency in Hollywood. People often want to define us by our work, which is an important part of us but not really who we truly are.

I understand how important the food we eat is and this led me to study Ayurveda, the ancient Indian Science of health. It is very important to eat the food your being needs, not just what your taste buds are driving you towards. Through my own process, I realize that whatever I eat *affects* me tremendously. Food is fuel and it's important for me to eat healthy non-processed food whenever possible. I shop at Farmer's Markets and eat mostly organic foods. I have been a vegetarian for more than 12 years and enjoy the benefits of eating whole grains, vegetables, and fruits. I love to cook for others and share my knowledge of vegetarianism and Ayurvedic cooking. Eating well is one of the keys to good health and maintaining a sustained energy at work.

The other key to maintaining my good health and great attitude is meditation. Sitting down in front of my altar and meditating in the mornings always inspires me and helps me to cope with the normal stresses of daily life. Those few precious moments are mine and they help to calm my entire being. I feel more positive in my daily life, I feel happier within and I am more connected to everyone else. I can't really explain it logically, it's just a feeling inside of me. The more disciplined I get with my meditations, the more I can see the results in my day-to-day life. I have also learned to not beat myself up if I don't meditate; instead, I focus on getting back to my practice, finding that time and space to connect once again.

I believe the process of meditating often intimidates many people. They are unclear of what to do, or afraid that they will have to sit endlessly without any thoughts, which are all misconceptions of the art of meditation. The important thing is to clear the time and space to sit with yourself and allow that experience of connection.

Trust me, it will serve you during your day more than you can imagine. It will allow you to be more centered, more balanced and ultimately to be healthier and clearer during the day. Meditating gives you the capacity to withstand the pressures of business life. I recommend the combination of meditation, exercise, and eating well to make you a well-balanced person at work and outside of it. It will give you the energy and strength you need to keep your life enjoyable and balanced.

It may appear that I want you to do this little thing for me, but it is actually for you. Read the next part and see where you spend your time and energy—see if you want some of that back.

How many hours a day do you waste on the Web, not doing research but gaming? Shopping? YouTube? Facebook or the newest social sites? Porn sites or dating sites? Be truthful and really look at it. I'm not asking you to give up these pursuits, but I do want your honest awareness directed at how much time you devote to these. See how much time slips by in these pursuits. So how much, truth telling?

Be aware of why we say we are resisting something and why we really are not usually in the same ballpark. What is your real resistance to success?

We all fear change because it is the unknown. "What we have, we know." I know it is challenging and maybe scary to think that you can't reward yourself or keep yourself awake and clear without your usual crutches. I know you may hate exercise, or perhaps you may think meditation is for everyone else but you. Or you may love being a night owl because you get to have time for yourself. There may be a million and ten other reasons not to do any one of these. Trust me, I know them all because at one time, I have used these excuses or heard them all. I am a disciplined person, but I still didn't want to get up at 3am; I didn't want to give up dairy; I didn't want to be accountable. I've been where you are. Are you ready? I tell you; I am more successful on every level now than when I was posting all that resistance against change. Embrace this wholeheartedly (that is such a great saying). You can always go back to what you have now, but you can't go forward if you aren't willing to do something different. Choose to focus on the questions that pertain to your most important of the 6 areas, then list how you will work to empower this area.

The Accountability Factor starts here—take a few moments each day and answer the following questions. Choose to focus on the questions that pertain to the first area you have chosen to put your energy into empowering.

Daily checklist for successful people:

Did I start my day meditating/sitting quietly for 11 minutes?

Did I take a cold shower this morning?

Did I eat something that fed my entire being for breakfast?

Did I do some exercise/stretching/yoga before work?

Did I take breaks at work today to connect with myself?

Did I take 3-minute exercise breaks during the day?

Did I remember to drink lots of water and tea today?

Did I stop to eat lunch calmly?

Did I pay attention to how I felt after my lunch?

Did I take a quick nap if I needed it?

Did I remember to breathe and use my breathing to calm me down during the day?

Did I take a shower at the end of my day, especially if it was a challenging day?

Did I do a couple of exercises to clear my day before sleep?

Did I end my day with meditation or an inspirational poem or both?

This is an assessment tool. If you are focusing on one area, please answer the ones that apply to that area. If you are taking on all of your blank pages in these 6 areas, then answer them all. These assessment questions are trackers. If you look at this list, you may wonder where your day has gone and why you have not done any, or hardly any of them. Count yourself among the millions who just are more in love with how things are than how they really want them to be. ***Don't worry, you can change camps anytime you want as soon as you are ready to go through the pain to get what you say you want.*** It does take courage and lots of desire and drive to make changes. Habits are truly what they are—bits of things that grow the more you do them. GOOD NEWS is you can consciously create new habits that feed your life any time you desire. *You don't need to get rid of bad habits, just feed the ones that truly feed you.*

You will have the time you deserve to do the things you really want to be doing with that extra time if you follow your choice by feeding it each day.

If you know what you truly want to be doing and you have passion, very little will stand between you and that goal—even death. This program is not for the average person—it is geared for the fearless extraordinary person. I hope that is you.

I have gone to enough conferences to know that there are those who buy all the programs. There are also those who focus only on the one purchase they have made, while there are those who try to do both. The first group who buys everything keeps hoping something will work. But they aren't doing the work—what they really want is a magic pill. The second group is the dive deep group. They pick one program and work through it, and then see if there are other components they may need to support them on their quest. The third group is so stretched out trying to do it all but they succeed at nothing. Be the middle group.

Plod through this if you must, and just stick with it. If you do this, you will feel amazing in so many areas of your life. THIS IS MY PROMISE TO YOU. You have been given the starting tools. If you want more support in creating new habits that serve you in-stead of you serving them, join me for a free 15-minute connect and see if we can work more in-depth or just answer a question you may have together (https://gurutej.com/bookings), then I can support you personally in creating sustainable habits. True Success starts here. Focus on your 3 things.

I will not try and tell you there is only one way but I want to tell you a story about a good way—the way that gives you a chance to embrace yourself as an excellent *hueman* being who wants to excel.

One time, I was at lunch with a couple of friends when Robin, who owns her own talent agency in Beverly Hills, volunteered what she needs to do to make her feel good in her life. This is what she said in the presence of someone who was a total stranger to her. "I have to do the things that support me or I'm just not right with my life." When my agent friend was asked by the other person with us at lunch what she meant, Robin said, "I have to meditate each morning, and eat right each day, and this means almost being vegan for me. The idea of exercise is not that great for me but taking a walk each day really makes me feel so much better. It is harder when I travel—trust me. I just came from New Orleans and they really looked at me as if I was weird when I asked for something that wasn't fried and had no meat in it. But I would rather get weird looks than feel totally weird." This could be your lucky day when you come to realize that you want to trade in old habits for ones that can serve you.

TESTIMONIAL:

Mark Burton, President of IndieVest Pictures

This has been a challenging time in my business for a variety of reasons. I was around meditation my whole life but I definitely got off the meditation track a while ago. I'm happy to say my wife was the great influence in my life. I saw what her daily meditation was doing for her, so first, I just joined her in her meditation. When I wanted something more specific for my situation, I reached out to Gurutej. I wrote her and she got back to me with the perfect thought for this phase of my business life. She gave me the meditation to do ***to make everything seem small*** so it won't overwhelm me or my business.

I must tell you the profound effect this meditation has had on me. What a journey I have embarked upon since I began regularly meditating this year. "Priceless, oh, priceless!" I do feel that there has been a real shift and I am ready to move into the New Year and seize the changes that are underway. I told Gurutej, "You have picked such transformative meditations for me and I was wondering if you could suggest a new meditation since I finished my last 40-day meditation that you recommended. I appreciate your guidance and support so much."

THE CYCLE OF OUR HABITS

Are you ready to free up your inner knowingness from the lies you tell yourself to keep doing the same things that don't serve you anymore?

I will tell you a secret: we are cyclical beings. There are inner cycles as well as outer ones. You have internal cycles that are specific to you. Then there are the larger ones and we call them external cycles like the seasons of the year. We eat differently in the summer than in the winter (hopefully). We eat differently when we exercise and when we don't (again, hopefully). We breathe differently when we exercise (our breathing gets longer and deeper). Some cycles serve us better. Understanding cycles can be a distinct advantage. It would help to know how long it takes to break an old habit and how long it takes to create a new one. Some say it takes only 21 days to break a habit, but I have studied habits and what it takes to shift them. I have to agree that it takes at least 40 consecutive days to break an old habit. You get to build new muscles on every level. It's what you are going towards, not what you are leaving behind. Build that muscle. What helps is having the right tools to reach our goal. Why do you care about breaking old habits and creating new ones? Usually, it's not for fun but rather because you want something different in your life.

I want to give you the tools to create a healthy day that gives you more energy and more hours to do what you really want to be doing. Here is a secret that will seem obvious once you see it.

Are you ready to understand how long it takes to break an old habit and how long it takes to create a new one? The creation of a new habit takes longer than it takes to break one. Breaking the habit is only one-third of the way. Here is the numeric breakdown: 40 days to break a habit and 120 to create a new one.

How do I know this? As I stated, I have studied this ancient yogic science and here is more proof—our cycles tell us this. Look around you. Look at the clues of 40s in our history—40 days of Lent, Advent, and Ramadan, 40 weeks of gestation for each child, 40 years spent by the Jews in the desert. If you want to break an old habit, then I suggest you try this program for 40 consecutive days for best results. Remember, anything is good but for a full effect, you need 40 consecutive days. To replace the old habit with new ones; to really cement the habit and make it yours, you need to repeat the new habit for 120 days. Scientists say that when you let go of an old pattern, there is a rivulet in your brain that literally disappears. When you create a new one, you actually get a new rivulet or pathway in your brain. Our brain is an amazing thing—it registers what we do and is far more elastic than we know.

120 days! I can hear you thinking, "That is so long—4 months!" Yet we all know to get really good at anything, we get to practice it over and over. Ask any great athlete, entrepreneur, etc.

Now think of all the things you have done for 120 days and have not blinked an eye. Let me start the list for you: brushing your teeth, breathing, eating, sleeping, and speaking. Now, your turn! What have you done for 40 days that you haven't even thought about?

1.

2.

3.

Now list three things you have consciously done for at least 120 days—things you chose and did:

1.

2.

3.

"Focus on changing ONE thing first. One habit. When you take on your keystone habit then all your other habits will align to make that succeed."

—from *The Power of Habits* **by** *Charles Duhigg.*

What do you have as a goal that you really want? Are your actions aligning to that desire, that goal? Let me share a story from Duhigg's *The Power of Habits*:

There was this Major in the army who was stationed in Kufa, Iraq. There were so many protests and they were not of the non-violent variety. So, he studied videos of these uprisings. He studied them for patterns. Then he went to the mayor of Kufa and asked his help in something very strange. He said, "Can you just ban the food vendors from the square, just for the next 40 days?" The mayor agreed. Then they waited. Within the next few days, people congregated in the square. It started getting loud and slightly angry with loud speeches just as he had viewed them on the videos, then people started looking around for the kebab vendors. They were hungry. When none appeared, the crowd just started disbanding. Without the food to keep them going, it all fell apart because this crowd had a pattern. He used his understanding of habits to undo a rebellion in Kufa. When the pattern was broken, they made other choices.

What do you need to break the cycle? If you want to fall asleep and wake up feeling good, pay attention to your nighttime patterns and what you automatically do when you get up. If you want to make exercise easy, create triggers, make it a routine. Once you see habits, you can go to work on them.

The book, *The Power of Habits*, gives us this idea of how habits are formed. First there is the cue that tells your brain to go into automatic mode and choose which habit to use. Then there is the routine which can be emotional, mental or physical. Finally, there is the reward which helps the brain figure out if this particular cue is worth remembering for the future. As Charles Duhigg says, "Habits aren't destiny. Habits can be changed, ignored or replaced. When a habit emerges, the brain stops fully participating in decision making. It stops working so hard or diverts focus to other tasks. So, unless you fight a habit, unless you find new routines, the same pattern will unfold automatically."

Our cravings and desires are what make cue-reward work. All advertisers know this, and they want you to crave their products. *I want you to crave your greatness, your potential for extra time and all that can mean.* There is that connection we get with the reward and that is being able to see something that we equate with the reward.

What if every time you saw your shower, you thought of the burst of energy you got taking your cold shower each morning? Or seeing a green juice and remembering how clear and alive it made you feel? What is the simple cue or signal that you are going to create to link to your one thing or habit? When your brain starts craving the reward, e.g., how great you feel after a yoga class or a run, this is the reward (the feeling), and this is what creates a habit.

Our unconscious habits can be corrected as well—eating from your child's plate, wiping your child's nose, correcting your spouse, etc. You can create new cues and rewards for not doing those unconscious things.

You can use this basic formula:

Create a cue (seeing your running shoes or your meditation spot as soon as you wake up.)

Make a list of your triggers.

Pick your area from the 6 above—the one you will work on and then fearlessly ask these questions:

What happens right before you want that food, that cigarette, that drink, that wanting to go to sleep or not wake up? Do a fearless inventory of yourself.

What is the reward you get from the habit you currently have in place? There is always a reward. You just have to decide if it is the best reward.

To change a habit, you must also find an alternative routine. Two researchers, Megan Oaten and Ken Change, enlisted 24 people ages 18-50 in an exercise program. They were interested to see if it would stick and how it would affect other parts of the participants' lives. These people were self-professed couch potatoes. After 2 months, they drank less, ate less junk food, spent more hours doing the things they loved, and became less depressed. YOU only have to change one area and it will effect the others; that's a promise, love.

Here are some other things to consider in order to create new habits:

Distraction from the reward. When you do something for some 21 days (as some may say, but I say 40 days), you have broken an old habit. To embrace your new love (habit), it takes 120 to build that muscle of a new habit. You want the reward that this new habit gives you. If you don't get into it yet, you will keep trying. If it hasn't become a pattern, you can be distracted—taken off your track, diverted. If you are 10 days into a meditation and you like it fine but then miss a day, no big deal. The big deal is getting back on track because it isn't yet part of your habits. If you have done a meditation for 90 or 120 days and you feel the difference it creates in your life, you will fight for that time to meditate.

The belief factor. For something to really change, you must believe this change is feasible. Belief is a non-provable but real factor in changing habits. Even scientists will confirm this. Thus, when you proclaim to a group or at least one other person that you will do something, it makes it more tangible, more real.

Keystone habits. Tricky science says these habits are the ones that will give you "small wins." They help other habits to flourish. They create new structures and establish cultures where change becomes contagious to the other habits you have in place. Starbucks says the most important aspect of creating new habits is willpower. This, they say, is the single most important keystone habit for individual success in anything.

Olympic habits. These are the ones that create wins because they are routine. Olympic Gold Medalist Michael Phelps' coach told him to watch a particular video over and over, until the video was in Michael Phelps' head. The visualization of his strokes, his turns, etc. is what made all these habits, and truly great habits saved him in an Olympic race where his goggles leaked and fogged. He had done it all blind. He knew the number of strokes it took him to finish a lap because he had visualized this all before.

What is your video? Write it down in detail. Yes! Right here in the book. or on your phone or… just get it down.

Experiment and try different things until you find something that works. Start recording your one thing or habit: How many times this week you meditated, both morning and night. Or if it is food, write down everything you eat. Put it in your phone since they are always with us. It can become contagious and spill over to creating awareness into other areas of your life. Just allow that awareness to come and then run after it.

When you force yourself to do your one thing or habit, you are changing how you think. Todd Heather, researcher at Dartmouth, says, "People get better at regulating their impulses. They learn how to distract themselves from temptation. Once you get into the willpower groove, your brain becomes practiced at helping you focus on your goal. You can choose to get overwhelmed or you can choose to say, 'Give it to me. I want more hours in my days.' What do I need to do? I want more hours and I want my health." You can get more hours by jacking yourself on drugs and caffeine, but you will guarantee yourself a shortened, unhealthy, anxious, or depressed life.

Are you ready to start?

This can be totally fun and a real adventure. Let's start where we should start. Waking up has everything to do with going to sleep—when and how you go to sleep. So that's where we shall start.

THE TRUTH ABOUT SLEEP

Sleep is such an interesting topic, isn't it? There are those who find it easy to sleep and those who don't and the latter expands to those challenged with either going to sleep or staying asleep (or maybe both). The "experts" are saying 8 hours of sleep is needed like a hard and fast rule. I'm saying there are a lot of variables. Let's peer into this sleep thing, yes. I know some people love, love to sleep and others almost fear it because they have such a hard time successfully embracing this thing called sleep.

In this chapter, you will get some great guidelines to hopefully join the sleep-with-ease group. For how you sleep determines how easily and happily you will awaken. No counting sheep or your bank account or going over your tomorrow's list involved. May this serve you well.

How you go to sleep matters. Getting ready for sleep in order to sleep well is an art and science.

Most people think it's really great if they shower, floss, and brush their teeth. These are all good and necessary, but we are just going to go a couple of steps further. These are real quickies:

a. Run your feet in cold water, or take a cold shower, then briskly towel-dry yourself till you are warm. This stimulates and clears the nerve endings in your feet. The soles of the feet, palms of the hands, and ears have nerve endings for all the organs in your body, thus there is hand and foot reflexology. You can clear the day and rejuvenate your nervous system by doing this one. Our nervous system takes such a beating with travel, stressors, children, worries, etc., so it's time to give it a little love and care.

b. Do a couple of exercises to relax your brain so it can get rejuvenated during sleep (which does not happen if you just fall exhausted into bed). Oddly, this first exercise can also be used to set your day because it both energizes and clears your magnetic field or the energy around your body which collects your dramas and traumas of the day and much more.

Claw Dance

Stand, with feet about two feet apart and your hands at your sides. Curl and tighten your fingers into long, claw-like hands.

Come up tall on your toes, swinging the arms out to the sides and up over the head in a big arc motion. Breathe in on the way up while saying "WHO." Next, squat down while the arms are swinging downward in a big arc. Exhale, saying "LA" on the way down.

Keep repeating these 2 steps for 1-2 minutes.

Single Nostril Breathing

The next one, I suggest that you share this with your spouse/partner, if you have one. Do the following exercise to help relax your Parasympathetic Nervous System. To sleep better at night, try breathing predominantly through your left nostril. Breathe slowly and evenly through your left nostril for 1-3 minutes. Just use any finger and block off your right nostril, thus allowing your left nostril to inhale and exhale both. Your left side is your feminine relaxing side. The Sympathetic Nervous System is the one that helps you to get up and go and rules when we are breathing predominantly through our right nostril. The right side is the masculine projective side (Reader's Digest description).

If you are already breathing predominantly through the left nostril, this will be really easy. If not, it usually takes about three minutes for the nostrils to switch from right to left and vice versa.

Sitali Breath, The Night or Relaxing Breath

Breathe out that stress.

Roll the tongue and stick it past your lips. If you cannot roll it, just make an "O" shape with your mouth, with the tongue protruding slightly beyond the lips.

Variations:

1. Lightly bite the tongue, inhale through the sides of the mouth; exhale through the nose.
2. Press the tongues onto the back of your front teeth. Allow the mouth to be slightly open and again inhale through the mouth over the tongue; exhale through the nose.

Breathe in deeply through the mouth. Exhale slowly through the nose, emptying the lungs all the way. Do this for 1-3 minutes.

Gift: The Sitali Breath is the best anti-stress breathing available. The breath passing over the wet tongue creates a cooling effect on your nervous system. Relax.

These few things will add a maximum of 7-8 minutes to your bedtime routine.

We all deal with sleepless nights from time to time—and yes, that includes experts like sleep doctors. Though having trouble falling (or staying) asleep can be distressing, it may be comforting to know that it's a common problem. According to a 2016 Consumer Reports survey, 68% of Americans struggle with sleep at least once a week.

"It's perfectly normal for all of us to have poor sleep or insomnia, but it becomes more of an issue for people that begin to become anxious and fixated on their sleep," Ruchir P. Patel, medical director of the Insomnia and Sleep Institute of Arizona, told HuffPost. "Remember, if you have a bad night here or there, it's normal. The more you stress about your sleep, the less it will return back towards normal."

I'm giving you this article to share sleep doctors' input. This is good and I'm with them on most of what they are sharing. My comments in parenthesis.

We asked sleep doctors to share the tips, tricks or other advice they follow when they personally have trouble sleeping. Some you can implement during the day as preventive measures; others you can try on a sleepless night.

1. They don't look at the clock.

> "When I am up in the middle of the night, I try my hardest not to stare at the clock (you can briefly check it don't use it as a gage for frustration) because clock watching is an easy way to have anxiety about the sleep loss and make the insomnia worse."
>
> —*Raj Dasgupta*, assistant professor of clinical medicine at the University of Southern California's Keck School of Medicine

2. They listen to soothing music.

"I will listen to Max Richter's 'Sleep' (I love the Relax app or chanting music) in the background, as this is very unique classical music where he worked with neuroscientists to create music utilizing tones that can help to relax the mind and assist with sleep. My wife and I also use it for our 15-month-old baby (since he was born) and [it] has helped him too."

—Patel

3. They try to find some enjoyment in quiet, cozy time in bed.

"I have learned to enjoy being awake in my bed. It's a pleasure to have quiet, comfortable time to reflect, plan, muse and simply be present. When you treat the situation as 'a living nightmare,' as one patient put it, you've already lost the battle.

—*W. Chris Winter*, president of Charlottesville Neurology and Sleep Medicine and author of "*The Sleep Solution*"

4. They remind themselves that they can always sleep in on the weekend (or some other time).

"It helps me to not worry that I did not sleep well last night or that I'm tired today."

—Patel

5. They don't look at their phone. ('tis true you get to judge your addiction to the little beast)

"I try my hardest not to grab my cellphone because once I open my phone, I'm suddenly answering emails, checking my Instagram and watching 'The Mandalorian' on Disney+."

—Dasgupta

6. They move to another room (moving to another room or even surface if you don't have a room) and read something dull.

"If ultimately I have a very difficult time initiating sleep (greater than 30 minutes), then getting out of bed and going to a different part of the house, using dim light and reading something boring on paper can help to distract my mind that I have not fallen asleep. Then, I return to bed as soon as I get drowsy."

—Patel

7. They try not to take a sleep aid if they can help it. (I recommend sleep aids 1 hour before sleep. See that section)

"I generally do not take anything for sleep as it is a slippery slope. It's very easy to become accustomed to taking something to 'help me sleep.'"

—Patel

8. But if they really need it, then take a small dose of melatonin (see range of herbs I recommend because if really challenged, something will work for a brief time then cease to be effective, so you want options).

"Rarely, I have taken melatonin (time release more effective) 3 to 5 mg as needed."

—Anupama Ramalingam, sleep medicine physician at the Insomnia and Sleep Institute of Arizona

9. They go outside in the morning (Look-at-Tshe-Sun meditation will help your pituitary set too),

"Bright light exposure can be accomplished by going for a walk or having breakfast outside on a sunny day, or by using a light therapy box for 30 minutes."

—*Stacey Gunn*, sleep medicine physician at the Insomnia and Sleep Institute of Arizona

10. They focus on resting rather than sleeping (in other words, relax, love; breath; see meditation).

"I know that restful meditation is a restorative process, much like sleep. Unlike sleep, resting is something I have complete control over. I choose to focus on what I can control."

—Winter

11. They make time for exercise during the day. (Yes, please!)

"To help with sleep, I get at least 20 minutes of exercise a day, whether it's just me on the elliptical at the local gym or the whole family taking a walk outside or at the park."

—Dasgupta

12. They tell themselves it's not a big deal if they have a night of poor sleep (yes).

"I keep reminding myself that it is OK if I don't get enough sleep tonight and that it's not the end of the world. This helps me to not get worked up, which in turn is going to take you longer to fall asleep. I also journal my thoughts, which helps."

—Ramalingam

13. They avoid sugar and caffeine later in the day (or, for some, after 9 am chocolate too).

"I try to eat smart before going to bed, which is not easy at all. I love junk food and Sour Patch Kids, but as a general rule, I try to avoid high carbohydrate meals and things with lots of sugar or caffeine."

—Dasgupta

14. They create their own bedtime rituals (see the section for great ideas to help with this).

"It starts two hours before bedtime and includes putting technology away, reading books to my children, (reading mystic poems) keeping the rooms quiet [and] cool, and turning down the lights. Happy and calm kids usually translate into better sleep for me."

—Dasgupta

MEDITATION 1. BEFORE GOING TO BED

Meditating with your spouse or partner (if you have one) will also make your love life MUCH MUCH better. You will be on the same page and connected.

Sit in an easy sitting position on a chair, on your bed, or on the floor. Make sure your back is straight, which means neither slumping nor hyper-extending your lower back.

If doing this with a partner, sit facing each other or back-to - back for physical or emotional support. If facing each other, hold hands. A good position is with each person with the left palm up and the right palm down on your partner's hand. Their left hand, palm down on your right, and your right hand palm down on their left hand.

Meditation 2. Creating Calm and Rejuvenating the Glands (more advanced)

Eyes are closed 9-10 seconds (called the drunken gaze; no booze involved, just inner drunkenness). Stare at the tip of the nose.

Hands are relaxed on the lap with both palms up, the left hand holding the right hand, the tips of the thumbs touching if doing this meditation alone; if with a partner, palms as suggested above.

Breathing: Inhale, sipping the breath in through the mouth to a slow 4 counts and exhale in 4 parts (snorts called Break Breath) through the nose, then hold the breath out for 8 counts.

Do this for at least 3 or 11 minutes at best. Then climb into bed clean, clear, relaxed, and ready for a great sleep of great connection.

Things that can Assist Your Sleep

I recommend using one of the following natural sleep remedies. They are easily available online (I use www.vitacost.com) or in your natural foods grocery store.

Rescue Remedy Sleep Spray by Bach *(One spray may be sufficient for you, but I have used a little more. Test it yourself.)*

Deep Sleep by Herbs Etc. *(capsules or liquid)*

MyoCalm *(I use this with Deep Sleep sometimes)*

Gaba Spray with Myocalm, or by itself.

Time Release Melatonin 5 mg. (some people need 10 mg.)

Organic CBD oil

Nutmeg (great in bedtime tea with milk substitute or Golden Milk; see below)

Warm, unsweetened tart cherry juice is not only very yummy but also helps regulate melatonin levels which control sleep patterns.

Take turmeric, which is also good for your joints, your heart and so much more. You can take pills—they are very good relaxers—or use it in your meals. I just found out you need black pepper with turmeric to make it work effectively. You can buy **Theracumin** which has the best black pepper and Turmeric combo to help your joints and more. You will have to play with these and find out what dosage you need, if any.

Yogi Herbal Tea makes a great tea called **Bedtime** (I love the Caramel flavored). Take with some almond (or rice, soy, hemp) milk in it. Can you tell I'm not into dairy? There are many reasons why. (This will be explored later.)

I have a friend who has tried everything on the market to sleep well—every herb, every vitamin and treatment, even surgery, but still no deep sleep. Then he found out that if he did a breathing exercise before bed and made this *Ayurvedic herbal combo*, he could sleep. Oh, and it works for me, too.

Golden Milk Recipe

Here is the recipe. All these ingredients are powdered. Mix a batch so it is ready whenever you want it. Make a big batch and keep it in a jar ready to help you sleep.

2 parts ginger root

2 parts turmeric root

1-part cardamom pods (green)

a pinch of nutmeg or more for each spoonful

1 part Freshly ground black pepper

Stir about ¼ teaspoon to a full tablespoon into a warm milk substitute. Start out slowly and see what works for you. Heat it until warm, and if you want, stir in a little honey—raw is always best—but don't cook it; just add it after you let it cool a bit. You only need a few ounces of fluid. If you can't be bothered with this, you can buy Theracumin which has the black pepper and Turmeric combo. Golden Milk tastes delicious and will help you sleep. For short sleeps, take a small amount; for catching up on sleep, take more. (This is especially great for travel and shifting to different time zones.)

My favorite milk substitutes are:

Coconut.

Unsweetened Vanilla (yummy and thick).

Almond. Organic, unsweetened; very sweet without added sweetener.

Rice. Always use organic, low sugar. Rice is one of the biggest GMO (Genetically Modified Organism) products.

Hemp. Great unless you have any form of herpes. It is very high in Arginine which will bring them out.

Soya. Make sure it is organic because almost all soy is GMO and many people have allergies to it.

Milk, dairy. If you can't give it up, make sure it is organic. Dairy is an acid food (more on that later). Let's just say acid makes you susceptible to disease. The best thing I have found for sleep, anxiety, and depression is **5HTP**, **Kava Tea** (or tincture) and **CBD oil**, and no, it's not a hallucinogen and they are not all created equal. I recommend NuLeaf because is the only truly organic one I have found, and hemp products are heavily sprayed. Don't want you to poison yourself in an attempt to be calmer and more relaxed. It is hemp oil, and yes, it has THC, but the one I'm recommending is legal in all 50 states and beyond. It is nanoparticulate, meaning you will utilize what you use because it is very assimilable and also liposomal, which means it will cross the blood-brain barrier—in short, be more effective. Just a dropperful or two (varies between us humans) is the usual dosage and will help you and your nervous system relax. This drink and meditation will get you into a deep calm sleep that will rejuvenate you but will not require 7+ hours of sleep.

Where You Sleep and What You Sleep on Matters

Sleeping in a clean uncluttered room, and not where you are working all day, is very important. I don't care if it means moving your office to Starbucks or the library.

Use clean sheets washed at least once a week, not once a month. You shed dead cells, you sweat, you do all sorts of things in bed. Would you wear a pair of underwear for as long as you sleep in your sheets without washing them? Get that image and act accordingly. You sleep better on clean sheets. Experiment.

Sleep with the right pillows. Go visit a mattress store or Bed, Bath & Beyond, and try pillows out. See what feels good to you—it's that important.

Knowing what kind of mattress works for you is also very important. They now make mattresses so that one partner can have the soft side and the other can have the firmer side. This is not a point for debate. It needs to work for you. ***Never take hand-me-down mattresses.*** Not only are they always over worn, they also carry lots of other peoples' embedded energy. People work out a lot of energy in bed through dreams and everything else. Start out with a fresh mattress, one without toxins. You can get these with a doctor's note saying that you have allergies. They now have a lot more natural mattress stores. In LA, there are a couple of places that have 2 natural bedding stores on the same block.

I was co-owner of a brass bed factory and we sold mattresses. I learned so much about people's attitudes when buying a mattress. I also saw what people did when testing a good mattress. It is very interesting to note the reasons people buy a mattress and not all of them have to do with sleeping. In our case, it is about sleeping, but test it for every activity. I know they are expensive, but not as expensive as your back going out on you. If you can't sit for any length of time without being in pain, then that won't work. If you can't do anything right now about your mattress, get a really good real sheepskin or wool mattress pad. It will help improve the bad situation, though it won't fix it.

Know if you need windows to be open or not, heat on or not, how many covers (or none at all), then negotiate with your partner if you have one. You can work around the fact that one person has more covers than the other. You may need separate rooms or you may adjust your internal heat factor (I will tell you how to do that with food). I always thought it would be sexy to have separate bedrooms and invite the other person in for the night, but I found that I really love sleeping with my husband and can't imagine separate rooms. Everyone is different, so find out what works for you. We do have very different body temperatures, so I know about negotiating the temp and cover thing. It can be done.

Turn off all electronic appliances in the bedroom 1 hour before sleeping to clear the magnetic field of the room. Or maybe have no TV in your bedroom—just maybe, yes!

Essential oils or good non-toxic incense will help you use the sense of smell to relax. Put 20 drops of lavender, chamomile, or marjoram in a small 2 oz. spray bottle and spray your bedroom. This also works wonders for your children's rooms and pillows. I love these oils (https://essentialthree.com) and I trust them. Go check out the blends too. Great sleep ones.

There Is a Secret to 10-minute Naps and The Best Hours for Sleeping and Naps

Google knows how to keep their people functioning. They have napping rooms. You don't need a napping room but just a floor, or a beanbag chair, or a thick yoga mat that is not in the middle of the workspace.

The following accessories are great to have but not absolutely necessary:

- phone with relax app
- an eye pillow to block out light (I recommend mine). See https://gurutej.com/product/chakra-pillows-bundle. This will explain all)

The best time for a nap is right after you have eaten a meal. Give yourself 10 minutes to digest your meal. If you eat heavy foods however, it will take much longer to digest.

The next best time is when you need a break or you are tired and experience brain drain. The best combo for rejuvenation is short meditation, exercise, and then a nap. You can do it in reverse order. You can take 15-20 minutes and you will come back rejuvenated and much more creative and productive.

There are some things you can do to get over brain overload in case you ever have it. Bowing is always good because it stimulates the flow of the cerebral spinal fluid and actually washes your brain out. Ready to be brainwashed?

Here are some things to try before or after your 10-minute nap. This also can help you before bed at night.

Four-Part Forward Bend

The brain is the fountain of youth.

Stand with legs wide open. The stiffer you feel, the wider you should spread the legs.

Bend over and place the fingertips on the floor right under the shoulders, with your head down. Bend the knees if necessary. When inhaling, lift trunk and head parallel to the floor. Exhale in four parts, lowering head and trunk towards the floor and in towards your legs in three equal parts.

Do this for 1-2 minutes.

Gift: This pumping motion activates the spinal fluid up and down the spine and clears your brain. This also allows your legs and spine to stretch. The enhanced breathing helps release overwhelming feelings or brain fatigue.

Spinal Rolls

Let the good times roll and relax.

Lie down on your back on a soft surface like a rug, yoga mat, or bed. Pull the knees toward the chest and wrap the arms around the legs. Roll back and forth. Inhale as you roll back, like doing a back somersault. Exhale as you roll forward until your feet touch the floor in front of you.

Do this for 1-3 minutes.

Gift: Motion and pressure points help calm the nervous system and reinforces the abdominal muscles.

Best Hours to Sleep and More

"Early to bed, early to rise makes a man/woman healthy, wealthy, and wise." Or does it? Perhaps with regard to rising early, yes. You may need more or less sleep. The "experts" now say 8 hours but that is if you don't get up and meditate. If you, do you need far less sleep, usually 4-6 hours only.

The hours you **need** to sleep should be from 10 pm to 4 am or it might be midnight to 4 am, or 11 pm to 3 am. (Remember you can go back to sleep after you meditate if you choose.) These are the most important sleeping hours because our circadian clocks want us to sleep when it is dark. It is the rhythm of our pituitary. That is what melatonin does—it works with resetting the pituitary gland, your master gland. (The pituitary gland is known as the master gland because it regulates the flow of hormones from the rest of the glands.) If you want to sleep less, sleeping at the peak hours is important. The best hours to get up and meditate are during what we call the ambrosial hours of 3 to 5 am.

The best way to stimulate the pituitary gland is to meditate. This, and natural light, are the things that set this gland best. The bone that covers the frontal lobe of your brain and your pituitary gland is the most porous bone in your body and is influenced by light. The pituitary gland sits behind this bone. Sunlight feeds and balances this gland—that is why people who live in climates where there is little to no sunlight in the winter suffer from Winter Depression. This is one of the reasons why it is best to sleep in a dark room so your pituitary gland knows it is night time and time to slow down on its stimulation of the other glands. Some people are more sensitive to light than others while sleeping. If you are this kind of person, wearing a sleep mask or having blackout curtains is a great idea.

If you are working across continents and nighttime is when you need to be working with your people in India, China, have a night job, etc., then you will need to sleep different hours. Work within your needs. Blackout curtains and eye masks will serve you well if you are in this category. I cannot tell you how many people have said to me that their best hours to work are 5 or 6 am when they can get to work before everyone else. By now, you probably know if you are a late-night worker or an early-morning worker. The thing with late night is that you usually have to jack yourself up with caffeine unless this is your peak time. Honor this if it works. I want you to get real energy and not use cheaters. So, proven best sleep hours, sadly for many, start at 10-10:30 pm, your time.

If you are in the habit of working overtime, you should also be aware that recent studies have shown a link to depression and excessive working hours. The reasons, as explained by *Web*MD, could be that longer work hours mean less time spent with loved ones and with attending to one's personal needs. This leads to lack of sleep and exercise, and an unhealthy diet.

If you work for a company, please have them study the Google model. We all know it is a massive success and there are very good reasons for this. They take care of their people. They have good food available. They have yoga classes and meditation classes and sleep rooms. They have exercise spaces. You don't need all this, but educate your company about how important breaks are in terms of better productivity (or bring me in to work with them and your entire company). Real productivity does not come from pushing and pushing some more. It comes from taking breaks, clearing your head, getting energy in your body, and then coming back to the project with new eyes and new ideas.

If You Can't Shut Your Mind Off, You Won't Sleep Well

You may think you are sleeping deeply but you will be:

a thrasher
a super epic dreamer
not sleeping well and you know it
sleepwalking/talking in your sleep
waking up often and not be able to go back to sleep easily

You will be busy because your mind is busy and your dream/sleep state will try to keep up with it. Meditation will help with calming the mind down and getting you to a relaxed state.

This is one of my favorite meditations to relax me and help even when I feel I can't get to sleep or go back to sleep

It's so simple you can even do it when you are lying in your bed.

Meditation to Relax You

Flip you tongue down your throat as if you were going to swallow it. (You won't.) This will cause you to put pressure on both your tongue and your pituitary, which is the gland that secrets the melatonin you need.

Slowly, very slowly and deeply, breath through your nose.

Do this for 3-22 minutes or till you feel calm and ready to let go and sleep.

Some people are afraid that meditation will make them flat and dull.

Truth be told, owning a calm mind is priceless—you become:

less reactive,
less emotional,
less angry,
more compassionate,
more connected,
more observant,
etc.

You are far from dull and you broaden your capacity to know what needs to be done and your ability to do it with ease is expanded. You will be able to see the talents of others and use them productively. There are so many pluses to meditation, I could go on and on.

What You Do During The Day Matters

This includes the food you eat, where you eat, your drink choices, your perspectives on events of the day, how you deal with the stressors along the way (your reactions and your capacity to release them), and truly, everything.

There is nothing that guarantees you will have no challenges, or that you will have lots of them on any given day. Some days you have such ease and flow and then… wow! I just spent 4 days trying to get my emails sent out. Was it easy to go through 4 days of frustration and turmoil? It's a big NO. I'm sure you have one or lots of these aggravations. We all do. During that time, if I had eaten bad food and just stayed plugged into my desk working away at it, I would have been fat, sick, frustrated, and feeling very overwhelmed. Instead I keep my yoga mat right by the desk and pop up into a headstand or handstand or some other favorite pose just to get the blood down in my head then back into my feet. The ultimate brainwash. You have these tools in this book and the flip charts. If you so choose, go here and say yes to visuals you can carry around with you. https://gurutej.com/store/yoga-flipcharts

What follows is a good exercise for clearing the head out during the day.

DESK BOWING

Self-reverence for a clear mind

Place fingertips on the edge of the desk or table with the arms extended.

Begin repetitive bowing. Quickly move head and torso up and down by bending the elbows. Inhale up, exhale down.

For added benefits, close the eyes and focus between the brows. Beware not to bang the head (padding the edge with a garment is recommended).

Do this for 1 minute.

Gift: This stimulates the lymph glands, enhances spinal fluid flow, and brings about instant mind clearing.

Should I eat before sleep? If "Yes," what?

I do best if I eat a few crackers or toast or a small spoonful of coconut mana. So, yes, carbs and oils. My husband doesn't do well eating after 7 pm. He eats early in the am and me not till 10 am usually. It's back to "thine own self be true."

Let's get some input from an expert, Tricia Williams, Culinary Nutritionist and Owner of Food Matters (*please note—I substituted green veggies for the chicken in this article.)

Your Sleep Problems Solved:

According to a recent poll by the National Sleep Foundation, 65 percent of Americans say they experience sleep problems, such as difficulty falling asleep, waking during the night and waking feeling groggy, at least a few times each week. The poll, which targeted 1,000 people across the country, found that nearly half, or 44 percent of those with problems, say they grapple with them almost every night. Every single night? That's just unacceptable.

At Food Matters NYC, my team and I have long promoted the idea of kitchen as cure—fridge as pharmacy. In other words, instead of rushing off to ask your doctor for an Ambien prescription, I suggest heading straight for your kitchen.

Bottom line: You can solve your sleep problem by "treating" it nutritionally. I do it for my private clients every day, and the results are staggering. But if you (like most people), scrape together your meals without the help of a private chef, don't worry. I'll show you how to get the same great results on your own.

To help support a night of restful, restorative sleep, the most important thing that you can do is fill your dinner plate with foods that contain tryptophan. Tryptophan is an amino acid that makes serotonin—a neurotransmitter that slows down nerve traffic to the brain. And slowing down that traffic is precisely what you need in order to drift off.

No doubt, you've been hearing about this essential ingredient for years, perhaps on Thanksgiving, when stuffed family members, dozing off before dark, blamed the tryptophan in the turkey for their sudden, crushing fatigue. But to really get the benefits of tryptophan, you'll need to pair it with foods that work in concert with it—maximizing its positive effects.

One of these "assistants" is calcium, which helps the brain turn tryptophan into melatonin. An ideal evening meal would include complex carbohydrates, protein and some calcium. Not too complicated, right? But just in case you're wondering what that would look like on your dinner plate. Here's a sample menu, to help you visualize:

- *(Green Veggies) with Quinoa**
- *Kale and Shiitake Mushroom Salad with a Honey Almond Soy Vinaigrette.*

Here's how I came up with this menu: First of all, quinoa is both a complete protein and a complex carbohydrate. What's more, it's high in magnesium, which aids in the fight against leg cramps (a common sleep foe) and helps the body process calcium more efficiently. Kale scores high marks across the board. It's an excellent non-dairy source of calcium and delivers an additional dose of magnesium.

And though many nutritionists will tell you not to eat late at night, there's new science that says it's okay to break that rule. In fact, if you are plagued by insomnia, you might actually benefit from a late-night snack. Only caveat: We're not talking Ben and Jerry's.

I've come up with what I think is the perfect before-bed treat. It works so well that one of my clients has taken to calling it nature's Xanax. I call it Kudzu Pudding. Kudzu is a root vegetable that has a very calming effect on the body. You can buy it in powder form at your local health food store.

What's great about using it in cooking is that it acts as a thickening agent and is really easy to work with. I like to cook it with low-

glycemic fruits like apples. It's a delicious desert and an effective tool toward winding down.

Here's how to make Kudzu Pudding: Start with 1 cup of applesauce in a small saucepan. Add 1 tablespoon of kudzu root powder. Bring the mixture to a boil, stirring constantly until the sauce thickens—about 3 minutes. That's it! Allow it to cool before eating—then make a beeline for bed.

So, if you're not getting enough shuteye, why not head to your kitchen cabinet rather than your medicine cabinet? You'll be surprised how a few tweaks to your diet can help you get your Zs.

(*https://www.huffingtonpost.com/entry/sleep-foods_b_1292099.html*)

The Truth About Waking Up Properly

Do you look forward to getting up or dread it?

What do you want to do first things in the morning?

Are you passionate about it or dreading it?

One will call you to action and the other will make you want to hit your snooze button over and over as long as you can.

This is where you "get to" set your day. Hitting that snooze button several times does not give you space to wake up properly.

This is where you get to look at self-uplifting or self-defeating habits. If you decide that you want something, then your actions need to support that desire. Otherwise, your lips are saying one thing but your whole being is not going along with your lips.

Doing these simple things at the beginning of your day sets your foundation for the rest of the day. It's called Awakening Success.

These simple exercises do something more. They actually bring all the parts of your being together in one place all aligned with each other.

We have 10 bodies:

Soul or **Essence Body**

Protective or **Negative Mind**

Positive or **Projective Mind**

Neutral Mind (the one you want to develop)

Physical Body (Your housing for all these bodies; get to know yours)

Arc Line (the glow you see around saints' heads). Yes, you also have one. Everyone has one, ear to ear. Women have two Arc Lines; the other is nipple-to-nipple of the breast. This is why men never want to lie to women as they will always know.

Aura or **Magnetic Field** (The energy around your body that both brings energy in and releases energy. Did you think only our appliances had one?)

Pranic Body (Your breath or energetic body which also connects your more etheric bodies to you.)

Subtle Body or the **Akashic Record Keeper** (This is really your know-it-all body. It does know all and is the storehouse for all you know. You get to it through the Pranic or breath body.)

Radiant Body (We all have this body and when developed, a person is glowing, and yes, radiant.)

This is an entire course in itself on the 10 bodies. I just want you to know that you are so much more than you think. We think we are body and mind, and sometimes soul. We are more than that and there are real ways to work with each of these bodies.

When all your bodies are working for you instead of against each other, you can get so much more done with so much more joy in life. That is what this program and book can do for you—get all your parts working together. When you are an effective organism, then the parts of your life can also be effective and powerful.

When we feel drained, there is no radiance, no energy, and no hope. My partner's Nanna calls people who are like this "no hopers." Such a great term, but I don't want you joining that group now or ever. If you have the desire to excel in your business, your personal life, and your life as a *hueman* being, this program will support that. You can truly have all, and yes, be radiant, alive, and kind.

Kindness comes from having enough yourself. It's hard to be kind from a dried-up well stream. When you are not nourished, you become the opposite of kind. You get reactive, depressed, and anxious. Pick your personal favorite.

So, these little things that will take you all of about 15-30 minutes each morning can mean the difference in experiencing the joy and spark in your life, or not. If you choose to go for that snooze button, it is really pressing the button that says, "No, I don't deserve to excel in all areas of my life. I will just do what I have always done."

Remember, the known is so much safer than the unknown even if it makes us miserable. If you'd rather stick with the known, then just pass this book on to someone who will use it. But if you want to excel in all areas of your life and enjoy all of it, then go forth and be willing to change to get what you want for your life, both business and personal, to a place of both success and joy.

So here is how to start your mornings (read this the night before the day starts so you can do it in the morning):

Stretching in bed before you get up for 5 minutes will realign your brain, making you ready for the day.

It will balance both hemispheres if you do diagonal stretches. Here are some really easy ones.

1. ***Lateral Stretches.*** Lie on your back and have both arms above your head. If you don't have enough room, then bend the arms, still in the position with arms up. Now stretch your left leg and right arm away from the body on the inhale, then exhale, relax, and switch sides. Do this 5-10 times. Later, stretches.

If you need energy, do this kind of breathing.

It is called **Breath of Fire or BOF.** It gives you energy and gets you up and moving. It will take away your anxiety and awaken your state of mind.

2. **BOF** is a rapid, continuous, diaphragmatic breath. Try this lying on your back, putting one hand on your diaphragm (in front of your body, located right under your ribcage). Then through your nose (or mouth, if you feel stuffed up), begin this rapid breathing; make it much more rapid than you think it should be. Feel your abdominal and diaphragm areas expand upward and outward to both sides of your body on the inhale and contract on the exhale. Many people are reverse breathers, meaning they expand when exhaling and contract on the inhale (you don't want to do that). It is retrainable to breath correctly with expansion on the inhale and contraction on the exhale. You won't hyperventilate while on your back. If you feel lightheaded, pull up on your Sex organs and Rectum.

3. **Do a couple of cat stretches.** On your back, keep the left leg straight and bend the right leg, bringing it over the left, then taking the left arm and helping the right knee towards the bed. Look to the right. Take a few breaths or more if time permits, and switch sides. This wakes up all the glands and massages the internal organs.

Then, sit up and massage your feet for 15 seconds each, or longer if you want. Sometimes, your feet are not ready for duty when they hit the floor, so squeeze them a little and wake up all the 72,000 meridians in them to wake up the flow of energy to all your internal organs.

NOW YOU ARE READY FOR THE BATHROOM

Brush your teeth and, for your tongue, use a tongue scraper or your toothbrush to get rid of the toxins that have collected on it during the night. Use a Neti Pot (which looks like a fairy watering pot or some torture tool). You can find one at your local health food store, on Amazon, and many other sites. You will put a pinch of high-quality sea salt in yours (I love to add a couple drops of colloidal silver tincture) and add warm water—not cold nor hot. (filtered water best) In your bathroom or any sink, turn your head to one side and place the little opening of the Neti pot gently at the tip of your upturned nostril and allow the water to move through and clear your sinuses. If it does not go in, try the other side and then come back to the first side. If this freaks you out, take a deep breath and then just do it. If your sinuses are really bad, stop and snort the water up your nose a few times and it will go in more deeply. Yes, you will have to blow it out afterwards but keep it in as long as you can.

This really always helps. If you fly a lot, it is a life saver. If you are prone to sinus infections, do this morning and night. Try it a couple of times and see if it can help. I have had people with sinus problems and ones with allergies that swear by this little device. It allows them to breathe and smell again. You don't need to help this process by taking a deep breath—it will just make you feel like you are drowning. If you want to take it in more deeply, sniff it up a couple of times after you have cleared your passages. Normally, you should allow it to flow in and out of your sinus cavity. I also do this with a product from Herbs Etc. called **Mouth Tonic.** It is really great for clearing the sinuses and doesn't dry them out like salt does.

Now, for the Rolls Royce of getting up.

TAKING A COLD SHOWER: OILING YOUR BODY FIRST WAKES YOU UP AND CLEARS YOUR HEAD

Now you are ready for the instant awakener. Brush your body with a loofah, body brush, or scrubby for a few seconds and then apply a light oil. You can get one that has the scent you like at your health food store or just use a good olive, sesame, or almond, then add your own essential oil of choice. I love Essential 3 https://essentialthree.com/product/hand-sanitizer-with-essential-oils/ food grade oils. Massage your entire body with this oil. It will keep you from having dry skin and will actually give you succulent skin. Who doesn't love feeling that on someone? It will also allow you to feel any abnormalities in your body long before they are detected by your doctor. Massaging your entire body with your hands each day keeps you aware of your body. It only takes a couple of minutes and it is great preventative medicine.

Cold Shower:

There is a right way to take a cold shower. This may sound funny, right? You might ask, "How could there be a right and wrong way?" Let's just say there is a more *effective* way. If this freaks you out, take a hot shower first, then a cold one which "gets" to be equally as long (3 minutes, hot; 3 minutes, cold).

Here is how you would want to take a cold shower:

Chant or sing enthusiastically.

Start with the tips of the extremities of your body—hands and feet first.

Rub the parts that are under the stream of water with your hands or your feet, depending on the part.

Make your heart area the last place to put under the cold water.

Don't forget your armpits—there are tons of lymph under your arms and around your breasts.

Enjoy. It is a wild and wonderful wakeup call and your nervous system will be so happy. This really is great for the nerve endings.

This was a yogi technique long before Wim Hoff, but he (bless him) has made it well known.

There is a VERY small number of people that can't, for some extreme health reasons, take a cold shower. Know if you are in this category, but that doesn't mean that those of you who have resistance should not do this, if it makes you feel uncomfortable. It means you have a health reason, love.

Cold water is not the same in all places. Los Angeles is not in the same class as Toronto, or the Arctic.

I have done cold showers all over the world and some truly make your entire being stand at attention; others are mildly cold. Be *grateful* that you can do it. It might come in handy if you travel to countries that have no hot water. Yes, they still exist. But most of all, it is for your nervous system—a huge boost to your health.

Next Is Meditation

Meditate before you start your day. You want something that will give you power for the day.

Try this meditation for activating your navel, which is your power connection to the world. Bring Earth and Ether together.

Meditation

Relax your hands on your knees, palms up, and touch your thumb and first finger together. This mudra or hand position is called Gyan Mudra or the position of wisdom.

You are going to make a sound here—yes, a sound you may not be familiar with.

Har Har Har means where you connect to everything—where finite and infinite connect. Do it by rolling your tongue over the roof of your mouth. If it is a rolled RRRRR, chant it with power so your navel moves in and out with each sound. This is called navel pumping.

Time: Do it for 11 minutes.

WHAT YOU EAT MATTERS

a. **Drink lemon water (preferably warm) before you eat, followed by tea, not coffee (Coffee is Hiroshima in a cup).**

b. ***Breakfast* is only the most important meal of the day for some people. Find out if you are one of these people, that is, if it is what will really fuel you for the day.**

Good brain food includes walnuts, oils, coconut oil, coconut manna, flax, avocados, and many other things. I'm not into fish oils because I'm vegan and because of the high mercury in fish. Make a protein drink with protein powder (my favorite is Sunwarrior Classic Plus Protein; all their protein powders are good) and add oil, nuts and berries in it.

The foods to avoid in the morning are sugars—yes, that means doughnuts, sweetened cereals, and straight fruit juice.

Foods to embrace are raw nuts. They all have different properties and gifts. Make your own mix of walnuts, Brazil nuts, hazelnuts, cashew, almonds (they need to be soaked overnight or toasted), pumpkin seeds, sunflower seeds, and some unsulfurized dried fruit (if that works for you and your digestion). Organic is always best in all products, but especially nuts.

A protein shake is another good fast option. All protein powders are not the same. My favorite is **Sunwarrior plus** chickpeas, amaranth, and fermented rice—yummy and works! I personally love the Vanilla and the Chocolate—plain for those who love it simple (http://www.sunwarrior.com).

They have other products I love as well. New Protein bars which rock 3g to 6g of sugar, and get this, 16g of protein. Isn't that fantastic!? I also love their Immune Shield and Liquid Light minerals. Check them out especially if Vegan or Vegetarian. You can add the minerals right into your shake.

Put your favorite milk substitute in the blender with the protein powder and some frozen or fresh fruit, plus greens. The low glycemic ones are the best. Blueberries, cherries, strawberries, blackberries, and raspberries help you stay healthy and not put on the weight. You can add chlorophyll or chlorella. Increasing the greens intake is always a good idea. Blend it in your "to go" cup and you are off.

Greens now have their own category of food—that is how powerful they are. It occupies more than a quarter of the Nutrition Food Plate. You can't get too much of greens. They are a nervous system saver. They are the stressor antidote. They make your nervous system go Ahhhh!

If you have watched the brilliant movie, *Forks over Knives*, *The Game Changers* or one of the many new ones coming out daily, you will find many reasons to shift to a plant-based diet at least more of the time. The movie *Forks over Knives* features the doctors who changed Bill Clinton from a meat-eating, junk-food-loving McDonald's poster boy to a plant-based-diet believer. He has now become one of the poster boys for plant-based diets.

THE TRUTH ABOUT THE FOODS YOU EAT

"What we include in our diet is as important as what we exclude."

-Dr. Dean Ornish.

What about the old saying you are what you eat? Do you see any truth to that? You will get to check in with you and see how food or what masquerades as food serves you, or maybe not.

You get to do that, should you choose deepen your discernment of what truly feeds you instead of judging yourself afterwards, which is so much fun right?

Please dive in.

YOU NEED TO EAT REAL FOODS, NOT THINGS MASQUERADING AS FOOD. THINK GREEN.

What do I mean by this? Well, anything with a shelf life of more than a week is not food. Anything that has words that you don't know the meaning of or can't say is not food. Anything that is fried or has high fructose corn syrup, aspartame, saccharine, or processed sugar in it is NOT FOOD. This also means ANY KIND OF SODA is not food.

Knowing this doesn't mean you are going to automatically stop eating these masqueraders, but I do want you to know two things: first, they are stealing your energy, and second, they are stealing your health.

These foods fake your body out because they bypass the brain receptors that say you are full. You will never feel satisfied nor full eating fake foods. You may feel bloated, you may feel gross, you may feel lots of things, but your body has no experience of having eaten. That's why we can eat so much chips or snack on "unfood."

You should want to help the planet as well as yourself go more and more plant based. Why? The number one contributor to greenhouse gas emissions is… are you ready for this? Cow farts and poop (not to mention the deforestation and land grabs to raise them and the emissions to transport them alive and then again, dead). Yes, it's true. So, if you eat less red meat, the law of supply and demand kicks in. If President Clinton (the premier junk food eater of all time whose idea of a good meal was Coca-Cola cake, and ribs) can go plant based, anyone, truly anyone, can do the same thing. That means you, and there's a super bonus that goes with it—inflammation in your body will go down, way down. And for most people, a huge boost in the E word. More Energy. Are you ready for that gift? Read this:

LBN-COMMENTARY BY JACKIE KELLER (FOUNDER OF NUTRIFIT):

We all know that eating a plant-based diet is healthier for our bodies than a meat-based one, right? And we know that meat production requires not only land but also energy and water. Aside from 16 pounds of wheat and up to 2,500 gallons of water being necessary to produce one pound of grain-fed beef, cattle production also consumes large amounts of fossil fuels—about a gallon of gasoline per pound of beef—and produces water pollution.

I suggest an alternative solution to the cost of producing meat that doesn't involve cultivating it in a Petri dish: how about we start eating more plants? In the words of Patrick Brown, a molecular biologist at Stanford University, "What you first need is a gateway drug for people to realize that all the things they love can be satisfied by plants.

Fruits and Vegetables Give Skin a Healthy Glow:

If just being healthier is not enough to motivate you to eat the recommended five fruits and vegetables a day, perhaps being more attractive is. According to new research, increased consumption of fruits and vegetables could contribute to a deepening of natural red and yellow skin coloration, which is thought to increase one's perceived attractiveness. The researchers believe that even small dietary changes could produce perceptible results. While experts have noted certain failures in the study's design, they agree that the findings could be beneficial anyway, since they encourage people to increase their fruit and vegetable intake. (Levine Breaking News, 3-12-2012)

A 2010 study found that people who drank both diet and regular soda gained more fat around their waist than people who did not drink any soda.

Warm Dr. Pepper was popular in the southern United States during the 1960s. This warm, syrupy drink was thought to help with colds and other ailments. It was also a popular drink at Christmas parties.

Drinking one soda a day can age you by as many as 4.6 additional years.

Most of the popular soda drinks today originated and flourished in pharmacies during the late 19th and 20th centuries. Early sodas were marketed as cure-alls for impotence, stomach pain, scurvy, alcoholism, opium addictions and nervousness.

Americans consume over 1.7 million tons of sugar each year from Coca-Cola alone. This is equivalent to 10.8 pounds of sugar per person, per year. Soda is the largest source of added sugar in American diets.

People who worked at soda fountains were called soda jerks because of the classic jerking motion used to pump soda water. But they made us all dependent jerks if we drank up.

In 2013, the Harvard School of Public Health revealed that 25,000 obesity-related American deaths from diabetes, cardiovascular disease, and cancer are linked to soda.

It would take 50 minutes of running or 5 miles of walking to burn the calories in a 20-ounce bottle of soda.

Coca-Cola uses over 300,000 tons of aluminum for its cans every year—just in the United States. This is 17.4% of the entire U.S. aluminum production. And all the soda companies want no responsibility for the recycling of their plastic and can pollution. So, they pollute us and our planet.

Another fun fact is what is around in most offices and maybe your desk drawer or sweets cabinet in your house. Just look at how you can unconsciously go mouth popping through that bowl of jellybeans or insert your favorite fake snack. The following story illustrates this.

Sodas' Disastrous Side Effects

A friend of mine had a sister whom we will call Betty. Betty was in her 40s when she had this illness that stumped doctors.

She just kept deteriorating at such a speed that even the doctors were freaked out. Yet, they had no answers for what was creating this collapse. Betty was then confined to a wheelchair but was adamant about staying with her plan to keep her scheduled trip around the world. The doctors said, "Yes, sure. Go. You may not make it back, but go, who knows?

Her sister, whom we will call Jane, was then listening to a program on the radio which started documenting all the disastrous side effects of sodas with all its chemicals mixed with sugar, high fructose corn sugars or worse. They said if there was only one thing you would cut out of your diet, make it sodas. Well, Jane's sister, Betty, started out each day with a soda and kept a steady stream of them coming all day long.

Jane called her and said, "Are you still drinking sodas?" Betty answered that, yes, she had just opened one at that moment. Jane then told her of the program and that it stated how drinking soda created all sorts of undetected diseases in the body that the medical profession had no way of detecting.

Jane then asked her sister if she was willing to try not to drink soda during the trip and see if she'd feel better at the end of it. Betty was willing to try anything, so she agreed. She started out her world trip in a wheelchair and ended it back at her doctor's office walking in. When her amazed doctor asked her if it was her trip that cured her, she said, "Yes, but what really made me well was when I stopped drinking sodas." She told them she had changed nothing else in her diet but just that one thing. They were flabbergasted that soda could have created all these kinds of disastrous inflammatory and debilitating effects. She is a living testimonial of the effects of soda and saying NO to soda.

This Article Is Even Scarier or Let's Say Hopefully More Awakening

A New Zealand woman dies from drinking more than two gallons of soda a day

(https://grist.org/food/new-zealand-woman-dies-of-a-coke-overdose-and-we-mean-the-drink/)

An even more disturbing story is that of a 30-year-old woman from New Zealand whose habit of drinking large amounts of soda (10 liters or 2.2 gallons a day) may have contributed to her death. The woman died from cardiac arrhythmia and also suffered from low potassium levels and caffeine toxicity. Her family narrated that drinking a popular soda was the first thing she would do in the morning and the last thing she would do at night. They haven't seen her drink anything else and the fridge was always full of her soda. When she ran out of it, she would "get the shakes, withdrawal symptoms, be angry, on edge, and snappy."

Bless Mayor Bloomberg for attempting to eliminate 32oz sodas in NY. Why? They are huge contributors to diabetes, large weight gains, heart problems, and much more. These foods I call cheaters are helping to collapse our health care system and even more, YOUR health. They will suck you dry of energy and make you fat and lethargic and you pay for them on both ends.

Energy Drinks

While caffeine content in soda is regulated by the FDA (the limit is 71 mg per 12 ounces), energy drinks have been largely unregulated since entering the beverage market. Many of these drinks use artificial sweeteners and are high in empty calories that lead to weight gain. Red Bull, Rolling Thunder, Rock Star, 5-hour Energy drink, and all the other "sports drinks" may give you the JOLT for a few hours, but they steal from your adrenals, your heart, your nervous system, and your breath. What do I mean by this? When you are amped up, you are not breathing deeply. Shallow breathing gives you less oxygen, less patience, less blood going to your head to clear out the old ideas and give you space for new ones. The sugar strips the myelin sheath off your nerve endings which makes you very intolerant. Caffeine is a bad lover. It jacks you up and then lets you down hard. Also constricts your muscles, thus its called a vasoconstrictor. Plus, it makes you very uptight. It's great if you are taking it to fight a migraine, but as a daily consumption item, it is so disastrous. These are what I call cheaters. Sugars and caffeine are the top two followed by the foods that weigh you down such as simple carbs, especially the white-flour carbs. Whole grains are good in small amounts, but most carbs will make you drowsy and thus will make you run for caffeine to awaken yourself. It becomes a round-robin of the food that makes you tired, then overriding it with the things that jack you up.

"Aside from its effect on the nervous system, new research has shown that the sugar in energy drinks causes tooth enamel damage that increases the risk for cavities and decay. The study, conducted by researchers of Southern Illinois University School of Dentistry, tested the levels of acidity in different energy and sports drinks and found that daily exposure to these drinks causes enamel damage in just a matter of days." (from the SIUSOD)

I will add that what it does to your nerves is even worse. It removes your myelin sheath. If that sounds scary, it should. It's the sheath that covers your nerve endings. Without it, you feel like an exposed nerve ending, because you are—jittery—not kind nor patient.

SUGAR

National Public Radio (NPR) had a 10-minute segment on how sugar is the cause of most inflammatory diseases. The doctor said refined sugar is the cause of so many diseases such as diabetes, many heart conditions, joint pains and arthritis, and many other inflammations. Please don't believe me; do the research yourself. While the issue of sugar addiction is still debatable, there is growing evidence convincing nutritionists and doctors that sugar addiction is real. Kathleen Des Maison, Ph.D., author of several best-selling health books, talks of how the over-consumption of sugar not only has a connection to weight gain but also to chronic depression. In Western societies like the U.S., sugar is so much a part of the diet that going without it could cause one to suffer moodiness, irritability, and other serious withdrawal symptoms, until you do some kind of detox. Do you know what the second most prescribed drug is in the US? Antidepressants.

Your sugar system is like a roller coaster that is out of control. This system will make you sick, crazy and dumb. It robs you of energy, clarity of thought and creativity while making you feel powerful and invincible—when all you have is an addiction to your unsuccessful habits. **It makes you a prime candidate for cancer, diabetes, and heart disease, just to name a few.**

I know all this may sound preachy or downright scary. I want to wake you up, not scare you. I want you to want what works for you, not against all parts of you with the exception of your blessed tongue who has undue voting rights.

My "Ask" of you, for you, and for the world.

Create successful food and beverage habits that will support and sustain you. Do you want to be there for your children's graduations and weddings, other future events that you hold precious? If yes, then, take care of yourself now.

EAT LIGHT, EASILY DIGESTIBLE FOODS

I realize that changing your palate is a huge step. Things that are good for you won't taste good to you if you have been mainlining junk foods. So, take it easy. Eat as often as you need in small portions. Yes, salads are at the top of the list. Make sure you have a salad dressing you really like (shop at health food stores because there are actually dressings without sugar that taste good, really good). Use spices and herbs to make things taste good to you. Also use sea salt instead of iodized salt. Salads are manly for all human beings because they support all parts of you. If they don't taste good to you, add a salad dressing you love, as long as it doesn't have sugar in it.

Super simple thing you can do is eat the dips you like—hopefully non-dairy ones, such as hummus, tahini, salsa, and guacamole with veggies or Romaine lettuce instead of chips. You will be full and clear. You can start out eating them with good crackers that don't have sugar or preservatives in them (please read labels); gluten-free is even batter. Mary's Gone has some great crackers.

ARTICLE ON FOOD AND OBESITY FROM HUFINGTON POST

Most of the overlap between the food world and the United Nations' mandate is about life-threatening hunger. The UN steps in to facilitate the distribution of food aid to those in desperate need. But a recent report from Olivier de Schutter, the UN's Special Reporter on the right to food, shows that when it comes to the UN's activities in the food world, feast can be just as important as famine.

"The right to food cannot be reduced to a right not to starve," the report explains. "States have a duty to protect the right to an adequate diet, in particular by regulating the food system.

In other words: letting the food industry create an environment that makes citizens unhealthy constitutes a violation of human rights. And the report argues that many Western governments have done just that by allowing "agri-food" businesses to sell fattening foods at low prices.

The report notes that over one billion people—one in seven alive today—are overweight or obese, a major factor behind the growth of many non-communicable diseases, like heart disease and diabetes. Such diseases have so far gone relatively unaddressed by the UN because they were not included in the organization's Millennium Development Goals, which have guided its activities since 2000.

But De Schutter's report isn't just a series of complaints. It provides possible solutions as well. Some are relatively uncontroversial. He argues governments should ban trans-fats, regulate the marketing of unhealthy foods and encourage the spread of farmers' markets and other access points for fresh produce.

The main suggestion of the report, though, is that governments implement taxes on soda and junk food—a perennially contentious topic. Specifically, the report recommends a 10 percent tax on soda and other foods "high in saturated fats, trans-fatty acids, sodium and sugar."

The report cites studies that have indicated that such a tax would reduce soda consumption by eight to 10 percent. And a recent analysis of soda taxes predicted that the measure would raise billions that could be dedicated to the support of health measures. Still, critics say such taxes are regressive and an unwelcome intrusion by an overactive nanny state. (Article in the Huffington Post 3-12-2012, "Junk Food Taxes Gain Advocate in United Nations")

Watch the movie *Raw for 30 Days* for free on YouTube. It is funny and deeply moving how these truly average Americans got off all their diabetes medications with food, meditation, yoga and group support in days.

Eat as Much Organic Food as Possible

Many grocery stores, including Costco and Walmart now carry fresh organic products at the same price and non-organic produce in their in their produce section, which is usually found at the front of the store. They have become tagged as the new darlings of the grocery section. Their way of making sure that shoppers keep coming is by putting organic products that are perceived as fresh and premium front and center of their stores.

You may also produce organic food in the comfort of your own home. Outside gardens, roof gardens, balcony gardens and even indoor herb garden kits are all available in many stores and online.

Here is an easy and yummy recipe of something nourishing that will be fun for you and your kids to do.

How to sprout any seed or bean:

Soak whatever seeds or beans you choose for 8 hours in a large glass jar, and then drain the water and rinse. Put a piece of cheese cloth or muslin over the mouth of the jar and fasten it with a rubber band and place the jar on its side. Any health food store carries a sprouting jar with all the parts you need if you don't want to collect these few parts on your own. Rinse and drain the contents 1-2 times a day. By day 2 or 3, you will have little tails coming out of the seeds or legumes. This means they are ready to eat or be blended with other things. You can use them to make great hummus. You can even cook the beans after sprouting, so you don't have gas problems.

Raw Hummus Recipe

You can use a food processor or Vitamix. You will need:

a few handfuls of sprouted chickpeas

1tbsp. or more of tahini

bunch of parsley

bunch of cilantro

a few pitted green olives

salt

lemon juice

organic cold-pressed olive oil

pinch of chipotle powder

1-2 cloves of garlic

a little water

You can alter this to taste. Sometimes, I leave out an ingredient or only use part of one ingredient, like less parsley or olives or add sun-dried tomatoes and grilled peppers. It depends, so give it the taste test.

Here are a few more recipes I love and are easy to do.

Green Dip

1 bunch parsley

1 bunch cilantro

2 cloves garlic or more to taste

1/2 lemon juice

sea salt

pepper, ground to taste

Olive oil

A little H_2O

Blend and voila! Amazing dip. It's a hit every time. I serve it to anyone on any type of diet. Also, you can add more olive oil and a little vinegar—your choice on what kind and blend as a salad dressing.

Vegan Dark Chocolate Mousse

This healthier 4-ingredient decadent mousse is dairy-free, rich and creamy, and easy to make!

Author: Anjali Lalani

Prep Time: 15 mins.

Total Time: 15 mins.

Course: dessert, snack

Cuisine: gluten-free, Italian, vegan

Keyword: easy chocolate mousse, vegan mousse

Servings: 3 servings

Calories: 381 kcal

Ingredients:

- 15 oz full-fat unsweetened coconut milk such as Thai Kitchen brand (see notes below)
- 4 tbsp. powdered sugar, or maple syrup (see notes below)
- 1/2 tsp. vanilla extract
- 3-4 tbsp. unsweetened dark chocolate cocoa powder (see notes below)
- toppings of choice - cocoa nibs, fresh fruit, nuts, shredded coconut (optional)

Instructions:

Place the can of full-fat coconut milk in the fridge overnight for at least 24 hours. Chill a small-medium size bowl in the freezer for a few minutes. Once you are ready to whip, remove the can from the fridge without shaking it. Slowly turn it upside down, and then open the can. You will see the liquid coconut milk (the part that did not solidify) will now be at the top of the can. Slowly pour out this liquid into a bowl or cup. This coconut milk liquid is not used in this recipe. You can save it to make a green smoothie.

Once you have poured out the liquid, the part that remains is the hardened coconut cream. Scoop out this coconut cream into the chilled bowl. (Note: do not use too large of a bowl, otherwise it will be difficult to whip.)

Using a stand mixer or an electric hand mixer, beat until fluffy. Start on low speed and work your way up to a higher speed until you start to see stiff peaks forming.

Halfway through, add the sweetener, vanilla extract, and cocoa powder and continue to beat.

Sprinkle with toppings if using any and serve immediately or store in the fridge in an air-tight container.

Recipe Notes

Coconut Milk:

Be sure to use full-fat coconut milk for this recipe, otherwise it will not work. They also sell coconut cream, which is the thickened part of the coconut milk once it's been refrigerated. This should work as well, but you will need to add more sweetener if using an entire can of coconut cream. If your can of coconut cream does not solidify, then the recipe will not work.

I have used Thai Kitchen Coconut Milk for this recipe many times and it's turned out great.

Sweetener:

You can use any sweetener, such as maple syrup or agave nectar. They taste good and work well.

Note that if you use liquid sweeteners, your end result may not be as fluffy or stiff because you are adding more liquid to the cream.

Unsweetened Dark Cocoa Powder:

I prefer organic Cacao powder. These are just a few of the easy wonderful recipes you can find on any vegan or vegetarian site. I must admit, I don't measure, so I give you permission to do the same if you're so inclined. You can make everything from smoothies to soups, to nut butters, to dips—you name it—with a heavy-duty blender like Vitamix. Look this up and don't be horrified by the price. But if you are, here are the top 10 possible replacement blenders:

http://www.blenderversus.com/cheaper-vitamix-alternative/

Make food prep fun and easy. Crock pots (make sure there's no lead in them) can enable you to cook while not home. Cook in batches, so when you are just too tired to cook or just want to open your fridge hoping some good fairy has left something for you or your family to eat, it will be true.

The vitamin will take the place of your blender and your food processor. It is one of the best investments you will make for your health. It is so fun and it can blend anything in my opinion. Just make sure you have enough fluid when blending. They say it can process wood chips (I believe it). Vitamix mixers are near-indestructible. Oh, and they're great for Raw Chocolate Mousse.

What You Drink During The Day Makes a Huge Difference in Your True Sleep

Caffeine, if drunk after 9 am for most, can produce a more challenged sleep. This is not true for all but if you can drink a cup of coffee and go to sleep, it doesn't mean coffee is good for you, it means your adrenals are shot. Your adrenals sit atop your kidneys and they work for you, unless you drink lots of caffeine. Then they go on vacation 'cause you have told them with what you are drinking that you don't need them. You found something else to do to their job. I do know caffeine really disrupts sleep and acidifies the body, which leads to all sorts of health issues.

What caffeine does is it takes the place of your adrenals and creates those boosts when you need them, giving you an adrenaline-like shot when you are in challenging situations and keeping your energy up when necessary. Once your adrenals get the message that you don't need them because you are doing it with the caffeine, then they shut down. This is what happens when people say, "I can have a cup of coffee and just go to sleep." This is NOT a good thing because what you are really saying is that your adrenals are so turned off that nothing kicks them back into action.

I know that many get headaches when they stop drinking coffee. It is interesting how our body starts detoxing and this freaks us out. Detoxing is not fun. Drink DeTox teas by Yogi Tea (no, I own no stock) 3 cups a day and you will breeze right through caffeine detox. Oh, and it tastes good, too. They have several DeTox variations.

Need something sweet because you have a sweet tooth? Try fruit or get some THINK bars because they have zero or low sugar and high protein, so even half a bar will quell your sugar craving. Raw Chocolate bars are my favorite. Fearless Chocolate as is Honey Mama's in her many flavors, are yummy. They do not drop the caffeine nor sugar blast on you. It is smooth and gives you a little lift. You only need a couple of bites, not an entire bar. Again, Sunwarrior Bars… yesness.

Eating Small Meals Often Works Best For Most

For sustained energy, you need little fuel intakes that keep your engine going. Big and heavy meals take too much energy from you during the digesting process. The brain and intestinal tract are the most similar-looking non-dual organs (examples of dual organs are the kidneys and lungs). The gut biome is a huge science. The centric nervous system, which is the nervous system feeding the gut, your gut, gets 200-300 million impulses from the brain and sends 800-900 million impulses back to the brain. So, the majority of talking is gut to brain. This tells us something of their relationship with each other. When the digestion is working really hard, it takes energy from everywhere. But you feel it most in the brain, thus the drowsy, tired feeling that you often fight back with stimulants, keeping the cycle of addiction going.

So, give yourself the 40-day (time needed to break a habit) trial to see if this is really true. Don't believe me? Experience it. Test it out and see how you feel after your detox period. Eat lightly—mostly plants and healthy snacks. Use Original Cinnamon Spice Tea with a green tea bag if necessary, or Chai Green Tea by Yogi. These are hearty spice teas with body. You can't just go from coffee to Chamomile tea, not even herb teas (though they are lovely) in a snap. You may use them too, but when you want body and aroma, Chai without sugar is the answer.

Heads up: All Starbucks products are loaded with sugar. The vanilla they use has more sugar than a Coke.

To Eat or Not to Eat Before You Sleep—There Is No Absulute Law and Depends on Body Types. Find Out The Secrets.

Let me say this: if you are thin, then you probably need to eat something—not a turkey dinner, but just a little something, like a piece of whole-grain toast, a little cereal, or a few crackers, all with a little oil or coconut manna. My favorites are Mary's Gone Rosemary/Garlic or plain salted ones. Light and yet help the sleep fairy come for me.

THE TRUTH ABOUT MEDITATION

This is my love place. The soft downy nest amidst the challenges of life. The set and reset place. I trust you will find something that will serve you.

WE ALL HAVE A MONKEY MIND

A very intelligent friend told me that she didn't want to start meditating because she didn't think she would be good at it. Just so you know, NO ONE IS. We are not trained to be quiet or to appreciate quiet. We often look at meditation as if it were a game to win. If you want the winning hand, then you need to meditate. This will give you the edge you want, unclench your jaw, lower your blood pressure, improve your heart rate, and make you easier to be around.

There is this wonderful story of Dr. Bill Fair who was a renowned surgeon for prostate and kidney cancer at Sloan Kettering. He was not a warm, cozy guy, but very very brilliant. He totally pooh-poohed the crazy West Coast hippy approach until he got cancer and did all the traditional allopathic surgeries and chemotherapies with no great results. Then all his very bright peers told him there was nothing more they could do for him and gave him a few months to live. His wife talked him into a change of diet and meditation with a what-do-you-have-to-lose? approach. He was very reluctant but he had nowhere else to go, so he did it. And it changed him so much that people liked being around him. He was much better in his family relationships, and as a surgeon, he was still the best. He credits these changes with extending his life for years past his prognosis.

You don't need to get a life-threatening disease to start meditating. There is so much data now on how meditation improves everything from your capacity to think more clearly, to your relationships, to seeing what is really important in your life, and yes, even your sex life.

It is learning to refocus your priorities to get you back, not to the shell of a person that people like or fear or envy, but to the real essence of you. This may be the part you resist most but I will tell you, it also may be the part you are the most grateful for.

Here Is a Study Published in Psychology Today

Maybe meditation isn't so mysterious after all. Neuroscientists have found that practicing meditation shifts people's brain activity to different areas of the cortex—brain waves in the stress-prone right frontal cortex move to the calmer left frontal cortex. This mental shift decreases the negative effects of stress, mild depression and anxiety. There is also less activity in the amygdala, where the brain processes fear.

Jon Kabat-Zinn, Ph.D., of the University of Massachusetts Medical School, recorded the brain waves of stressed-out employees of a high-tech firm in Madison, Wisconsin. The subjects were split randomly into two groups, 25 people were asked to learn meditation over eight weeks, and the remaining 16 were left alone as a control group.

All participants had their brain waves scanned three times during the study: at the beginning of the experiment, when meditation lessons were completed eight weeks later, and four months after that. The researchers found that people who meditate showed a pronounced shift in activity to the left frontal lobe. In other words, they were calmer and happier than before.

(Colin Allen, The Benefits of Meditation, published on April 1, 2003) *http://www.psychologytoday.com/articles/200304/the-benefits-meditation*

I have found in the 50+ years I have taught yoga and meditation that breathing and sound are the fastest way to get to being calm and centered.

Your breathing, which is the easiest way to connect with your meditative mind during the day, is the most wonderful autonomic response. We will speak more about this in the breath segment. Breath is an integral part of meditation.

The great thing with meditation is that the effects last far beyond the time you are meditating. To back this up, here are some research studies by Richard Davidson, Ph.D., of the University of Wisconsin and Matthieu Richard, Ph.D., who studied the minds of regular meditators. They studied the brain's oscillations and found that the regions of the brain are said to synchronize. "On the scale of the brain, it is a phenomenon comparable to establishing coherence of the body's biological processes. Still better, they found this to last far beyond the actual meditation time." This is from the book, *Anticancer: A New Way of Life*, by David Servan-Schreiber, M.D., Ph.D., with an easily accessible research on meditation, great food tips, and more. This is not just for those who have cancer. It is also for those who don't want to have the disease. The preventative path is to see great input on the pros of meditation and be smart enough to say yes to it right now. Sorry, there are no cons. Well, our resistance and fear of not being "good" at it are cons, I guess, till you drop them.

This is something I have just come upon that has helped many who feel they can't meditate. It's good for physical pain and calms the mind. CBD Oil, and, yes, they are not all created equal. I like this company, Nuleaf, because they are the only one I found that is organic and sadly the marijuana industry uses a lot of chemicals in growing their crops. Thus, organic is really important. I am invested in this company's products because I trust their purity and the availability in your body because of the nanoparticles and the liposomal dispersion. https://nuleafnaturals.com/shop/

I encourage you to try it. I take it every morning before I meditate. and before bed. try it on your time off to see how it works on and for you.

MEDITATION CALMS THE MIND AND GIVES YOU SPACE FOR ANSWERS TO COME

Easily, it's the key to everything, especially to clearing your mind. Meditate first and last thing of each day and you will stretch time and space. You can also take it into your walking running exercise. Moving meditations can be a great way to start.

MEDITATE ON THIS: THE PRACTICE CAN HEAL YOU IN LESS THAN 11 HOURS. —STEPHAN A. SCHWARTZ

This work has documented a kind of deep "stillness" that affects the entire brain. When this occurs, the frontal and temporal lobe circuits—which track time and create self-awareness—seemingly disengage. The mind-body connection dissolves. These studies show us that the limbic system is responsible for assigning emotional values to persons, places, everything in our total life experience. Since the limbic system, among other things, regulates relaxation and ultimately controls the autonomic nervous system, heart rate, blood pressure, and metabolism, it produces both emotional and physiological effects when you react to a specific object, person or place. This is why your hair "stands on end," your skin "crawls," your stomach "lurches" or your heart "beats faster."

Because meditation affects the limbic system, developing the discipline allows one to become more volitionally in control of these responses. The practice has a calming effect that leaves us relaxed and physiologically more evenly-regulated. This, in turn, allows us to be coherently focused, because we are less distracted by our inner dialogue and emotions as well as our physiological responses. And this literally changes your brain.

(http://www.huffingtonpost.com/stephan-a-schwartz/effect%20of-meditation-on-brain-_b_941082.html)

The following meditation technique is one of the best. It is very simple you can do it almost anywhere.

Sipping Breath Meditation

Position:

Allow your lips to go loose and act as if you have a straw in your mouth and inhale through the straw and exhale through your nose (or stick your tongue through your lips and inhale from your mouth and exhale through your nose).

Do this for 3-11 minutes.

If you have had way too much caffeine and nothing will calm you down, this might do the job. You will have to do more than 3 minutes, but it will help stop the self-mutilation of the nervous system. I love this one when you feel confused or just need to see better options or possibilities.

Try This Next Meditation: When You Don't Know What to do

Before you start, put out in front of you what it is that you are worried, concerned, or just needing more clarity about, and then let it be. Answers will come, but not the usual ones.

This meditation is very easy, and at the same time, very complicated. Complicated because you have to stay focused and present or you will totally lose where you are (and That too is fine).

Position: Sit in an easy pose, or with your spine straight, on a chair.

Next, rest both hands, palms up on your lap, and let the right hand hold the left, with fingers pointing away from the body at angles, the right thumb crossing over the left thumb.

Breathe: Inhale slowly through the mouth and then exhale slowly, also through the mouth.

Inhale through the nose and exhale through nose.

Inhale through the mouth and exhale through the nose.

Inhale through the nose and exhale through the mouth, then repeat this sequence for a minimum of 3 minutes. 11 minutes is optimum.

End with 3 deep breaths. Hold the breath in at the completion of each inhale, then sit and listen to the answers.

Answers will come, allow them the space to do so.

Here Is Another For Instant Calm

Place each fingertip to tip with its partner finger on the opposite hand and very lightly rub your fingertips in little circles. Do this for 1-3 minutes. Your parasympathetic nervous system will literally purr.

I have taught many Hollywood people, some of the biggest agents in the business, who found this to be the best thing they have ever learned. They still do it years after I taught them.

Meditation Turns The Volume Down So That Clear Thinking Can Ensue

This is what all truly successful people can do—turn off the world, turn off the inner chaos. If you read their techniques, some just stop and sit in their favorite chair quietly while others do another kind of practice. But across the board, these techniques expand the breathing, which is a sign of inner freedom—meaning tension is not present. At any time, you can gauge your tension by how deep and relaxed your breathing is. They are in direct correlation with each other.

More stress means less breathing and more breathing means less stress. The clarity of your brain is absolutely related to the vastness of your breath.

I love love love this one more.

MEDITATION FOR PEACE AND CALM

Mudra or hand position: use of your palms touching. Fingers outstretched with thumbs (Mars) touching each other and pinky (Mercury) touching the opposite pinky. Other fingers outstretched like a lotus.

Eyes must be almost closed (9/10th closed) looking down into the palms.

Inhale through the mouth in a 4-part sipping breath, starting the inhale from as low in your belly as you can. Think of starting in your pubic bone. I love to use Sa (birth) Ta (life) Na (death, nothingness) Ma (rebirth) as you inhale.

Exhale through the nose in 4 equal parts starting the exhale high up in your chest and being totally without breath and your navel pulled back in when you have finished the exhale. Use the same mantra on the exhale.

Hold the breath out for 4 counts reciting Wahe Guru (The Divine ecstasy that connects you to everything) like you are wrapping yourself in a divine ribbon.

Start out with 3-7 minutes and build up to 11 or 22 minutes. I even do this one in my car (not with eyes closed, unless you want an accident). It will keep you calm and at peace. Use this for 1-3 minutes throughout your day.

All Highly Sucessful People Know How to Be Still Quickly

I travel a lot and at airports, I allow books to speak to me. I buy and read them on long flights. Sir Richard Branson's book, "*Screw It, Just Do It*" was one. He speaks of doing most of his great thinking on his houseboat. It is his meditation place. When his company went public and he had to answer to board members and lost his contemplation time, he freaked out and bought all the stock back. He needed to do it his way. That meant he needed time to be and do his type of meditation—stillness.

Try each of these things then pick the one that works best for you.

The Truth about Exercise

Without this one, love, you will become a rusty ole machine that will need to be abandoned long before its expiration date. If you are ready for some good maintenance tools that don't require hours a day… trek on.

Two Articles:

Renew Exercise

Your brain is a thinking organ that learns and grows by interacting with the world through perception and action. Mental stimulation improves brain function and actually protects against cognitive decline, as does physical exercise.

The human brain is able to continually adapt and rewire itself. Even in old age, it can grow new neurons. Severe mental decline is usually caused by disease, whereas most age-related losses in memory or motor skills simply result from inactivity and a lack of mental exercise and stimulation. In other words, use it or lose it.

Our Sedentary Society

Not too long ago, futurists envisioned humans evolving giant thumbs in response to a push-button world. They did not foresee humanity's real response to all its labor-saving conveniences—a sedentary, inactive society with a deteriorated vascular system and consequent decline in physical and mental health.

Nearly half of young people with ages 12 to 21 do not participate in vigorous physical activity on a regular basis. Fewer than one-in-four children report getting at least half an hour of any type of daily physical activity and do not attend any school physical education classes.

In June 2001, ABC News reported that school children spend 4.8 hours per day on the computer, watching TV, or playing video games. The impact of computers, video games, cuts on school funding and public apathy have combined to leave Illinois as the only state that still requires daily physical education in first through 12th grades. This is a far cry from the 1960s, when President John F. Kennedy made physical fitness a priority for Americans of all ages.

These sedentary tendencies represent a real health crisis, and, not just for couch potatoes. Deep vein thrombosis (DVT) occurs when blood circulation slows, allowing clots to form and then, eventually, break free, and causing death. DVT has been nicknamed 'economy class syndrome,' because airplane passengers who sit throughout a long flight in the close quarters of economy class have become victims of DVT.

(Both articles are from www.fi.edu; no author mentioned.)

We need to move during the day. Our bodies are designed with movable parts. Think of all the joints and what they tell you. In the 1950s, a study conducted on ticket takers for trains and conductors for trains showed that ticket takers who had to walk up and down the trains had 50% less risk for heart attacks. Is there anything more to say?

Find something you love and do it. Or, if you don't love to exercise, find someone who will do it with you that you want to spend time with. Go off your grid and try yoga or dance or do something totally different from what you have tried before.

Do you want stats about your heart, or your weight, or any other stats? Google it, but if you want the truth, exercise tells the body that it matters. It releases unwanted thoughts, stressors, opens up stale energy, and awakens endorphins. New studies have found that sitting at your desk for as little as 2 hours (without a break) is the equivalent of smoking a pack of cigarettes. We were designed to move. Our body, just like our pets, needs to be taken out and exercised. We hear great stories of how important exercise is, seeing how people who have been in accidents fight back to regain their bodies when they were told they would never walk again, much less run. Look at Lance Armstrong with his cancer. I'm sure you have examples in your mind of people who came back to life through great struggles to regain the use of their bodies, and we sometimes just ignore ours or get lazy. It wants your attention and it screams at you with aches and pains if you don't take them out on the open road and run them full throttle. Well, maybe not full throttle, but at least open it up. Your body is a gift—take care of it.

Did you know that you can get high cholesterol even if you eat well, or relatively well, but don't exercise or meditate? I know people to whom this has happened. A woman friend of mine was told by her doctor, "Your husband is giving you your high cholesterol." She knows she needs to both exercise and meditate but she loves her work to the exclusion of all else. This time, her doctor put the fear of heart attack in her and so she started 20 minutes on the treadmill with a chanting tape. This was her concession to meditation—at least she is moving and breathing.

So, start wherever you can. Acknowledge where you are, what you are resisting, then just pick something and do it despite the resistance. We can always find data to support our craziness. Don't let some serious illness have to chase you down to make you take action.

You get to choose between lots of medications in cocktails that will give you crazy side effects that sound like a horror movie, or you can move your booty. Have you ever listened to the commercials on TV when they list the side effects of the drugs? I wonder if people just blank those out because they want what is promised on the front end. It is really that simple. If you don't move it, you will be guaranteed a painful decline.

You may have lots of aches and pains but once you start moving, there is a good chance, with a high antioxidant diet, meditation, and exercise combo, your inflammations will go away. An alkaline, plant-based diet is also guaranteed to help with inflammations. With food, movement, and collective stillness, you will find the condition of your body will be lots more fun. If this is not something you resist, then move on to the next aspect of this course which you have the biggest challenge with.

I had an acquaintance that started a very successful online company and was glued to his desk. He put on 150 lbs. in one year. He had to sell the company and take a year off to recover. You can do this in your 20s and maybe part of your 30s, but after that, your body will really bite back. It gets mad at you for abandoning it. So, be good because it's like Santa watching you, wondering what you are going to do. He doesn't care if you are naughty or nice. He is just watching to see if you can outrun the reindeer. Here, the key word is DO.

What are you going to do? In most surgeries now, they get people up and moving the next day, even in hip replacements. They now realize that the body needs to move just as well as rest in order to heal.

If you are the person who needs support, choose something you like and get someone you like to do it with you.

If you like to do it alone, then do so, but make sure you have someone who supports you in doing it. Or join a team sport or dance or exercise group, be it runners, dancers, skaters, yogis—it doesn't matter.

Time is important to me. I am blessed to have hills around my house. I strap on my rollerblades and take off. I get 40 minutes of exercise rollerblading the hills. It is so fun and it clears my head out and I have developed great strong legs. I use yoga for my body and love it. I keep a yoga mat right by my desk... and swimming too, I love it. I am blessed with a saltwater lap pool right up the street that I didn't even know about for years till I really needed to find one after my back took me out. Oh, and by the way, I am 71 years old, but no one knows or even thinks of it when they see me zooming down the hills.

Have fun with life—it is not supposed to be drudgery. Incorporate what your needs are with where you live or work, with your pets, etc. Try something new—it keeps you younger. Use what you have and have fun. If you work alone all day, then join a club and be with people if that is something you need. Balance is getting what you need for all the different areas of your body and your life. If you have a partner, do something with them—sometimes doing what you love and sometimes doing what they love. You may both start to love each other's preferences. My husband joined me swimming to get ready for a trip to Tahiti we had planned and now he is ready to go 3 times/week even when I may not be able to. Exercising together puts you in sync with each other.

Find things you can do during the day at your desk (Examples will follow)

Don't wait for Tuesday, or Wednesday or... someday... to roll around so you can exercise. The moment to do something is when you are exhausted, frustrated, stuck, angry, anxious, depressed—need I say more? Do something now. Here are some things you can do at your desk https://gurutej.com/store/yoga-flipcharts. These are very simple tools to help you feel energized, connected and more whole.

This is one from the flips charts which offer. 1- to 3-minute- things you can do to release or rejuvenate yourself during the day or before bed.

HAND DUSTING

Cleanup time; get rid of the unwanted.

Sit or stand straight, hands in front of your heart, act like you are dusting a powdery substance off your hands. Do this quickly and breathe deeply!

Do this for 1-3 minutes.

Gift: Helps release mental pressure and worries, stimulates nerves in hands, creates calmness, makes you more energetic. You might look a little crazy but what the heck—you either want to look crazy or be crazy (anger counts as crazy)—you choose.

LASSO POSE

The energy wrangler.

Sit up straight, make fists with your thumbs tucked inside. Move arms wildly overhead in backward circles. All the while relax shoulders and engage the core to power the arms. Breathe!

Do this for 1-2 minutes.

Gift: Creates greater clarity and helps in being more receptive, increases heart rate, and very energizing.

Use tricks to get enough exercise, such as putting the yoga mat by your desk, running up and down the stairs, riding the bike to work, or walking the kids to school

Choose what you are going to do. I rollerblade and keep my yoga mat by my desk, and just go upside down to brainwash myself on a regular basis. Share with me what you are doing. What are your tricks? Let's support each other and expand our resources.

Stretching is the alter ego of hard workouts. Yoga helps you last longer at any sport you may love (factoid)

Try this for clearing your magnetic field (yes, we have them too).

Claw Dance

Two steps for a radiant, charismatic energy.

Curl up and tighten fingers into long claw-like hands. Come up tall on your toes, swinging the arms out to the sides and up over the head in a big arcing motion. Breath in on the way up while saying "WHO." Next, squat down while the arms are swinging downward in a big arc. Exhale, saying "LA" on the way down. Keep repeating these 2 steps.

Do this for 1-2 minutes.

Gift: Strengthens thighs and increases energy, replenishes your
energy field and expands awareness, helps clear the mind when facing challenges.

Love this beautiful body you were gifted with and take loving care of it. 'Tis the only one you get this time round. You asked for this unit, now you get to work with the care and maintenance of this wondrous unit. Read the maintenance instructions laid out here and add what you love.

THE TRUTH ABOUT BREATH—YOUR BREATH

Without this life hack, you cease to be here. Either you are barely breathing (existing) or you use your breath to shift your mental, physical, emotional, and, yes, soul states. Your Breath is #1 in your life tool kit.

THREE WAYS YOUR BREATHING CAN IMPROVE YOUR PRODUCTIVITY

"The Change Blog," by Christopher R. Edgar

While we're working, we often get so absorbed in our projects that we forget about our breathing. When this happens, particularly when we're under stress, we can lapse into restricted breathing—inhaling in short gasps, or shallowly into the upper chest. What we don't often realize is that how we breathe can deeply affect our efficiency and enjoyment in what we do. In this post, I'll offer three breathing techniques to help you stay focused and peaceful as you work.

1. Breathe through the Urge

Productivity writers often tell us to resist the urge to put off our work, but they usually don't offer much practical advice on how to do that. In my experience working with people on productivity issues, we often procrastinate when an uncomfortable thought or sensation—anxiety or anger, for instance—comes up as we're working, and we'd rather not experience it. We need, I think, some way to stay centered even when faced with those intense thoughts and feelings.

I've found that the best approach in these moments is to breathe deeply. Instead of turning away from your task to play FreeCell or instant-message with friends, hold your attention on what you're doing, and take full breaths until the difficult sensation passes away. If you start feeling anxious as you're working, for instance, breathe slowly and deeply until the anxiety dissipates.

When we keep breathing in the face of discomfort, often the difficult sensations we're feeling seem more manageable and less threatening. As Dr. Miriam Adahan puts it in Living with Difficult People, Including Yourself, *"When you keep breathing calmly or moving purposefully, your muscles will teach your brain that there is no real danger."*

2. Focus on Your Breathing

Zen meditators often concentrate on their breathing to stay alert and keep their minds from drifting into memories or concerns about the future. What I've found is that this technique isn't just useful for meditation—it also works great whenever we find ourselves getting distracted at work. We can focus on our breathing to bring our attention back to this moment, and to what we're doing.

Many meditation teachers explain why this technique works by observing that, whenever we focus our attention on what's happening in our bodies, our awareness naturally settles into the present. If I ask you to pay attention to your breathing, you probably won't start daydreaming about the way you used to breathe five years ago—you'll focus on the act and experience of breathing right now. When your attention returns to the present, the memories and worries that may have been bothering you fade into the background.

3. Notice How You Restrict Your Breathing

If you're feeling tense or uncomfortable as you're working, take a moment and notice how you're breathing. Are you breathing shallowly and rapidly into your throat or chest? Are you making breathing difficult by clenching any muscles? Some people I know check in with their bodies periodically as they're working and discover they're hardly breathing at all.

When we aren't allowing ourselves enough air, it's no surprise that working feels painful and worrisome. If you notice that you're breathing in a shallow or restricted way, see if you can slow and deepen your breaths, and let go of any tension that's cutting off the natural flow of your respiration. Working becomes easier and more fulfilling, I think you'll find, when you breathe fully as you do it."

—Jon Kabbet Zinn

http://www.thechangeblog.com/improve-your-productivity/

Our Breathing Is An Automatic Response

Your breath is your source of life. Without it, we cease to be in this body. With it we are alive. Fortunately, it is an automatic response, or we would be dead many times over for forgetting to, yes, breathe.

Breathing is our birthright—breathing well and knowing the kinds of breathing that will help you with your energy is something that would have been good to know from the time we were 4 years old or even sooner. Well, you will know soon.

Watch children breath and you can tell if they are secure or not. If they are reverse-breathing, you know that those around them don't know how to breath because they are stressed and the children pick that up automatically.

When you know how to adjust the rhythm of your breathing, you can literally control your moods, your reactions, and which system, the sympathetic or parasympathetic nervous system, needs to be stimulated at any given time. Know how to have the energy to go powerfully and how to slow down your rhythms to be receptive, to relax, or sleep. Do you want this practical knowledge? It will make your day, your life, and your mental, physical, and spiritual state so much better.

This is perhaps the most important section of this entire process. Our breathing not only tells us where we are in terms of our mental and emotional state, but it is also feeding our fears by robbing us of oxygen and life force, or it is adding to our life force and making us broader; more vast in our perspectives and possibilities. Because it is an automatic response, we don't actually die by forgetting to breathe—we just become smaller, more constricted, and tighter inside this body unit, and yes, that brings us to death's door much sooner.

I want you to know how to adjust the rhythms of your body through your breathing. The great thing about breathing is that you can do many of these anywhere—in a meeting, on the phone, driving, stuck in traffic, during a disagreement, before bed, when you get up. Being conscious of your breathing makes you conscious of so much more. Your breathing expands your lungs, and your lungs massage your heart; breathing brings in the oxygen needed by your blood and your blood carries oxygen everywhere including your brain. Rejuvenation! It's all because of breathing.

Breathing consciously is a privilege. When you learn how to breathe and you use these techniques, you will wonder why you never did this before. This expands your awareness more quickly and powerfully than anything else you can practice all day long.

There is one thing I would like you to do first. Lie down on your back. If this is not something you can do right away, do it when you get home. Put one hand on your diaphragm and the other on your chest. Now inhale the way you usually do. Check to see if the hands move up. When you inhale, you want to be expanding. You want to be getting bigger and fatter with the diaphragm and ribs expanding. If you are deflating, you are doing a reverse breath and want to shift this. Think of this: you are a balloon and when you inhale, you get to take in the entire universe with you. You expand. When you exhale, it is your chance to release the debris—the mental, physical, emotional toxins. You deflate the balloon, letting go. It's a great deal. I just want to make sure you know the basics of how to breathe. Truly, there are more people who reverse-breathe because they feel it isn't safe to breathe, to expand, to take up space. We learn this from those around us—it is not spoken of, just learned by proxy. Children watch and learn from what they are exposed to. This is your chance to reverse that learning and expand. It may feel really weird if you are a reverse-breather. Good news is I have taught hundreds, if not thousands, of reverse breathers to breathe the way their bodies were meant to breathe, and they all succeeded. Yes, some it took time to switch but all made it through. If you stay with it, your breathing will change faster than you think. The bonus gift is that you will be paying more attention to how you breathe. What a great deal that is.

There Is Not Just One Breath for All Occasions

There are so many types of breathing that help us in every situation in life—there is breathing for getting more energy, breathing for calming down, breathing to help us let go of anger, and breathing that breaks up overwhelming feelings. The yogis have known these for ages and now you can know them, too.

If you are a stress bunny, do the Four-Part Breath and Sitali Breath. If you are prone to anger, embrace Dragon Breath and Sipping Breath. (These are all explained below). If you never have enough energy, the Breath of Fire is your new best friend.

Read over them all; try them for a couple of minutes and practice the ones that resonate with you NOW. This will change because we are constantly changing beings. Remember to come back to these kinds of breathing and try new ones once every 40 days at least.

Some kinds of breathing are calming, some energizing, but all of them activate your awareness.

You have an entire army of breathing techniques to probe—go look at them and try them all often so that when you need them, they are in your data bank ready to remind you and work for you.

The following type of breathing is fun and is great to do with a partner, or in front of a mirror if you are taking yourself too seriously. Children love this and it is great to do when angry or upset. Can you imagine if it is totally acceptable to do this in offices as a positive way to channel anger and frustration? You can just notify those around that this isn't the best time to engage you when doing this exercise. This time will come. This is also so great to teach to kids, so they have positive ways to release their frustrations, and, yes, anger.

Dragon Breath

You will inhale in 4 equal parts preferably with your eyes closed (but not while driving) and exhale in one big breath out the mouth with your tongue stuck out as far as possible and eyes wide—very wide open. Do this 3-11 times for the best results.

Children love this one because it is playful. It also activates your pituitary gland which is the master gland in charge of all the others, getting the right messages from them and secreting what they need, all in the correct time. It also puts a pressure on your thalamus and hypothalamus. Don't believe me? Feel what happens in your throat while doing this. It also helps regulate your vagal tonality. This Breath is a great way to blow off steam. Be the dragon (but you don't need to incinerate anyone in the process). It is best to do this before speaking to someone we want to annihilate.

Having an Army of Breaths for All Situations Is Necessary

What is the best kind of breathing? There is no such thing—just the best breathing for the occasion. It is just like clothes—you don't wear a dress suit to a pool party, or a beach shorts outfit to a fancy dinner party unless you want to stand out. It is the same with breathing.

Make a list of yours. Or just get the flip charts. https://gurutej.com/store/yoga-flipcharts. They are all in there. Try to practice each of them at least once a day for 40 days so they will be part of your program and you can share them with others. Teaching them or sharing them with someone else is a great way to claim them on a deeper level.

Stress Robs You of Your Breath

Fact: Just check with your body for confirmation of this one.

Breathing Is The Fastest Road To Energy, Always. It's The Original Red Bull

When you are breathing shallowly or in distressed breathing mode, you are literally robbing your body of the oxygen it needs to do all the functions you are asking of it. Your brain gets fuzzy and cloudy and you get depressed or anxious. You overeat or don't eat at all to compensate for the lack of energy.

Your breathing is the source and I want you to have a totally new arsenal at your disposal. The only way they will become yours is if you practice them. Pick a couple and do them every day until they are second nature. Pick ones that are relevant to you.

When you need energy, you want things that are fast, like Breath of Fire. When you want to calm down, you need Sitali Breath or "O" Breath. When you need energy, breathing is your best friend, not caffeine.

Breath of Fire

Instant energy booster.

Breath of Fire is a continuous, rapid, diaphragmatic breathing done through the nose or mouth. Inhaling and exhaling are both done equally and evenly. Think of butterfly wings allowing the diaphragm to move easily and rhythmically.

For added benefits, pump with the abdomen while breathing: pull the navel inward as you exhale and release as you inhale.

Do this for 1-2 minutes.

Gift: Increases metabolism; excellent in rejuvenating and energizing.

Here Is Another De-stressor During the Day

Eye Calm

No strain. All gain.

Close the eyes and rub both hands together as quickly as possible for 60 seconds. Then cup the hands tightly over the eyes and open the eyes to look into the darkened palms.

Allow the eyes to relax in this position for a minute. For greater relaxation, roll the eyes (open or closed) a few times in each direction as well as up and down and then in a diagonal motion—upper right to lower left, then upper left to lower right.

Do this for 1-2 minutes.

Gift: Relaxes and re-energizes strained eyes. Helps you see more clearly.

What follows is another gift of breath.

FOUR-PART BREATH: MIRACLE BREATH

Break up with overwhelming feelings. Blow your stress away.

Sitting up straight, inhale through the mouth, filling your lungs in four equal parts—starting the breathing as low as possible in the abdomen. Exhale through the mouth in four equal parts, from the upper chest down. If you feel relaxed, you can do this exercise breathing through the nose.

Repeat the cycle for1-3 minutes.

Gift: Instantly reduces stress, so you may see a situation more objectively; helps dissolve overwhelming feelings; will make you more present with each situation in your life.

How can this be? You go get spaced out but then it makes you more present? Yes, you get to go far and wide so you can come back and become more linear and focused. This is the paradox of living here. You need to go wide and loose to get narrow and concise. You will never know if you don't try it. And doing something for one day or two weeks doesn't count—you need a straight, unbroken 40 days just to get started. As we said, 40 days is the time it takes to break an old habit. It takes 120 to create a new habit. These are consecutive days. You are creating a new groove in your brain. Be willing to care for you enough to say yes to consistency and caring.

Science has actually tracked the fact that habits have their own groove in our brains. When you let go of a habit, the groove vanishes slowly and new habits create new grooves. This gives new meaning to being groovy—it means having new insights.

This one is about experience only. I can talk until I'm blue, but experience is worth millions of words. The difference between prayer and meditation is that in prayer, you are calling the Infinite (or call it Energy or Essence if you want). In meditation, you are being receptive to the Infinite calling you. One is projective, the other receptive. We need them both. During these times, receptive is usually our weakest aspect. We get to balance the massive amount of projective time we have in each day with the ability to get receptive, yes!

SITALI BREATH

Breathe out that stress.

Take your tongue and press it into the back of your lower teeth. Close your mouth till slightly open. Inhale through the sides of your mouth calmly and slowly. Exhale slowly through your nose, emptying the lungs all the way.

Do this for 1-2 minutes.

Gift: The best anti-stress breathing available; the breath passing over the wet tongue creates a cooling effect.

O BREATH

Back to cool, calm

Make an "O" shape with your mouth. Inhale slowly and deeply through this "O." Exhale long and deep in the same way.

Do this for 1-2 minutes.

Gift: Helps regain calm and centeredness.

Your breathing can save you in each and every situation of your life. It is the real lifesaver. If you can slow your breath down, you can expand your awareness out. There is an old saying, "When you get your breath to one breath every 30 seconds, The Divine comes to talk with you; when you get your breath down to one breath per minute, The Divine comes to serve you." Whatever that might mean to you, it means power, any way you slice it. Control your breathing, control your emotions, your fears, your perceptions. It is the key to everything. Everyone can learn to do this. Your breath is free; use it well. When you can master your awareness of your breathing and the flow of it, you will master much in your life.

Taking Stock of Your New Life

So, to end this wonderful journey with you, I want you to go back and take stock.

Look at your answers to the questions in the beginning of the book.

Pick your strongest area and make a decision on what three things from that area you will implement to have the feeling of true success.

If you already feel successful and want to tackle what is un-nurtured, pick your weakest area to feed that starving part of your being first.

Write these three commitments down and please include meditation and breathing as a part of whatever area you choose. Just a few minutes, yes! Build your calm, clear, aware self.

Take exercise/breath breaks in your day so you don't feel broken, brittle, or on the overwhelmed track.

You have great tools but they are only great if you claim them and use them. Make them into habits that will serve you forever.

Practice means you repeat these tools over and over each day until they effortlessly become part of what sustains and supports you in claiming your totally successful life.

Connection to yourself will give you the capacity to be connected to others. We are on a relational planet. Everything is about our relationships, whether they work or not. These techniques can make them work.

FNAL TIP: SUMMATION

When you are really tired, comb your hair with a wooden comb, take a cold shower, do some breathing exercises, and have a green juice. You will come back to power right then. If not, try a power nap or some exercises by your desk. Avoid the cheaters—sugar, caffeine, cigarettes, alcohol, or food. Go for what will really feed you.

When captured behind our desk or sedentary, we tend to eat more because we are burning brain cells but not body energy, nor calories. Choose what you eat wisely.

If you have made it this far, chances are you are serious, so go implement "your" 3 things in your daily life.

You will enjoy your life more, have more energy, work with less effort, and be much more successful in all areas of your life, not just one.

Thank you and bless you always on this journey into true success to regain your true life. Allow your new habits to serve you love. Our habits serve us or take us down. May yours be great servants of your true essence and all your being.

www.ingramcontent.com/pod-product-compliance
Lightning Source LLC
LaVergne TN
LVHW050936210926
848482LV00013B/636
* 9 7 9 8 2 9 5 4 0 1 9 9 2 *